D1442245

Outdoor Masonry Projects
& Porch Construction

Complete Handyman's Library™
Handyman Club of America
Minneapolis, Minnesota

Published in 1996 by
Handyman Club of America
12301 Whitewater Drive
Minnetonka, Minnesota 55343

Published by arrangement with Cy DeCosse Incorporated
ISBN 0-86573-677-4

Printed on American paper by
R. R. Donnelley & Sons Co.
99 98 97 96 / 5 4 3 2 1

CREDITS:
Created by: The Editors of Cy DeCosse Incorporated
and the staff of the Handyman Club of America
in cooperation with Black & Decker. **BLACK&DECKER**
is a trademark of Black & Decker (US), Incorporated
and is used under license.

Handyman Club of America:
    Book Marketing Manager: Cal Franklin
    Book Marketing Coordinator: Jay McNaughton

Contents

Introduction 5
Building with Masonry Products 6
    Planning & Designing Masonry Projects 8
    Masonry Tools 10
    Safety & Disposal 12
Building with Concrete 14
    Concrete Basics 16
    Preparing the Project Site 20
    Placing Concrete 26
    Finishing & Curing Concrete 30
    Building with Concrete: A Step-by-Step Overview 32
    Walkways 34
    Exposed-aggregate Patio 40
    Installing Tiled Patios 46
    How to Lay Patio Tile 53
    Concrete Steps 60
    Footing 66
    Sealing & Maintaining Concrete 70
Building with Brick & Block 72
    Brick & Block Basics 78
    Mixing & Throwing Mortar 82
    Laying Brick & Block 84
    Building with Brick & Block: A Step-by-Step Overview.....................92
    Brick-paver Step Landing with Planter Option 94
    Brick Veneer 100
    Cleaning & Painting Brick & Block 106
Planning a Porch 108
    Working with Inspectors & Building Codes 112
    Working Safely 114
    Tools & Materials 116
Porch-building Projects 118
    Building Porches 120
    Building & Installing Porch Railings 148
    Building Wood Porch Steps 152
Index 158

Introduction

The patio and the porch are traditional focal points for outdoor living. Few elements of your home will conjure up crisper images than the recollection of a holiday picnic with the family out on the patio or of a breezy front porch on a perfect summer evening. Sidewalks and steps, and brick landings, planters, and veneers also contribute to the atmosphere of your outdoor lifestyle. *Outdoor Masonry Projects & Porch Construction* gives you complete instructions for building a variety of these projects, which can be constructed as shown or easily modified to fit your needs. Along with the projects, you will find information on basic tool use, material handling, and building techniques.

By using masonry as a building material, you can create patios, walls, planters, and other projects that will last for decades. And, in most cases, you can create these projects at a lower cost with masonry than with any other building material.

The first section of *Outdoor Masonry Projects & Porch Construction* guides you through the tools and materials that you will need to work with concrete, brick, and block. You also learn how to design and plan masonry projects, observing the relevant building code. Information on working safely with these materials as well as on proper disposal is also provided.

The next two sections show you, step-by-step, how to build structures with concrete and brick and block. Poured concrete projects are made by building forms, usually from wood, and filling them with concrete. You can shape the material to meet your needs and the special dimensions your project may demand. Brick, block, and other building materials provide an attractive appearance and texture, and can be stacked in just about any pattern or style you like. In both sections, you learn how to maintain what you have built.

The final two sections of the book provide you with all the necessary information to plan and build a porch, including railings and steps. You learn about the tools and materials you need, plus how to work with inspectors and the appropriate building codes. Each element of building a porch is demonstrated through complete step-by-step instructions.

Masonry projects or building a porch are often major jobs that usually require some help. The instructions and information in *Outdoor Masonry Projects & Porch Construction* will guide you through all aspects of making these projects part of your home. It will help you determine what you can do yourself and what you may need professional assistance with—and how to work with professional designers and contractors to make certain you get exactly what you want.

NOTICE TO READERS

This book provides useful instructions, but we cannot anticipate all of your working conditions or the characteristics of your materials and tools. For safety, you should use caution, care, and good judgment when following the procedures described in this book. Consider your own skill level and the instructions and safety precautions associated with the various tools and materials shown. Neither the publisher nor Black & Decker® can assume responsibility for any damage to property or injury to persons as a result of misuse of the information provided.

The instructions in this book conform to "The Uniform Plumbing Code," "The National Electrical Code Reference Book," and "The Uniform Building Code" current at the time of its original publication. Consult your local Building Department for information on building permits, codes, and other laws as they apply to your project.

Building with Masonry Products

Masonry products can be used to build projects that are almost unlimited in shape, size, color, or texture. The ability of poured concrete and brick or block to conform to just about any job is one of the greatest benefits of building with masonry products.

On the next few pages, you will find helpful photographs of several masonry projects that can be built by most do-it-yourselfers. Take note of the unique design features in each project, and consider how you can apply these ideas to your own yard or house.

It is extremely important that your masonry building project meets local building code requirements—always consult a building inspector before committing yourself to a plan or design.

This section shows:

• Planning & Designing Masonry Projects (pages 8 to 9)
• Masonry Tools (pages 10 to 11)
• Safety & Disposal (pages 12 to 13)

Bring out the best features of your yard. The owner of this home designed and built these decorative fence columns to match the brick-paver walkway. This enhanced the design of the walkway, while complementing the brick veneer around the entry area of the house.

Adapt materials to match your space. A winding walkway, steps, and stoop like those shown above, can be designed to fit just about any landscape. Use isolation joints between structural elements to break large projects into smaller working sections, making the project more manageable.

Compare samples of masonry building products to see how they will look in your building site. Sample cards, whole bricks, and concrete pavers can be ob- tained from most brick yards. Pay special attention to existing masonry colors, styles, and textures around your home to see if the combination is pleasing.

Planning & Designing Masonry Projects

There are two basic stages to planning and design- ing any building project. First, gather creative ideas to help plan a project that is both practical and attractive. Second, apply the basic standards of masonry construction to your idea to create a detailed plan that is structurally sound and com- plies with local building codes.

Good masonry design considers size and scale, location, slope and drainage, reinforcement, material selection, and appearance. All of the tips and information on the following pages are designed to help familiarize you with these de- sign elements.

If you do not have much masonry experience, start with simple, stand-alone projects. Building a small garden wall or pouring a short backyard walkway are good first-time projects. Repairing existing structures also provides valuable expe- rience for future building projects.

One of the best ways to acquire design ideas is simply to take casual walks through your neighbor- hood. Bring along a notebook and record any comments about masonry structures or surfaces you see. There is no better way to know what a de- sign will look like than to see a similar design in completed form.

Tips for Designing a Masonry Project

Test layouts for planned projects by using a rope or hose to outline proposed project areas. This will help you make decisions about size, scale, and shape. For curved walkways, use spacers between the borders of the project site to maintain an accurate, even width.

Make detailed plan drawings. By using graph paper to create scaled drawings of the planned project, you can eliminate design flaws and make better estimates of building material requirements. In some cases, local codes may require you to obtain a permit before beginning the project, and scaled plan drawings usually are needed for permit approval. Always check with your local building department early in the planning process.

Common Masonry Projects Around the Home

Project	Level of Difficulty	Special Considerations
Walkways & Sidewalks	Basic to Moderate	Square corners are simple to create; curves and angles complicate the project; walkways are subject to codes that govern size, reinforcement, location, and allowable materials; often built adjacent to other permanent structures, creating a need for isolation joints.
Patios & Large Slabs	Moderate to Advanced	Require extensive use of control joints and isolation joints; volume of concrete requires very efficient placing and finishing techniques; can be broken down into several small projects with the use of permanent forms; establishing slope can be tricky.
Brick & Block Walls	Basic to Advanced	Simple garden-type walls are easy to build; project complexity increases with size, complexity of stacking pattern, and need for reinforcement; walls over 3 ft. tall generally require a frost footing; wall caps or free-end pillars may be needed.
Poured Concrete Walls	Advanced	Require complicated form bracing and extensive reinforcement; usually are load-bearing, and subject to extensive code standards and may require inspection by building departments.
Driveways/ Garage Floors	Advanced	Subbase preparation and grading are very important and time-consuming; can require specialty tools, like bull floats, to accommodate high volume of concrete; high-strength concrete often required; garage floors often steel-floated for a hard, semi-glossy surface.
Functional/Decorative projects (e.g., brick barbecue or planter, casting concrete pavers)	Basic to Moderate	Simple projects with few structural requirements; good projects for beginners.

Tools for mixing concrete and for site preparation include: a sturdy wheelbarrow (A) with a minimum capacity of 6 cubic ft.; a power concrete mixer (B) for larger poured concrete projects (more than ½ cubic yard); a masonry hoe (C); a mortar box (D) for mixing mortar and small amounts of concrete; a square-end spade (E) for removing sod by hand, excavating, and for settling poured concrete; a sod cutter (F) for stripping larger areas of sod for reuse; and a tamper (G) for compacting the building site and subbase.

Masonry Tools

To work effectively with concrete, brick and block, and other masonry products, you will need to buy or rent some special-purpose tools.

Trowels, floats, edgers, and jointers are hand tools used to place, shape, and finish concrete and mortar. Bricksets and cold chisels are used to cut and fit brick and block. Equip your circular saw and power drill with cutting discs designed for use with concrete and brick to convert them into special-purpose masonry tools.

Successfully mixing concrete and mortar also depends on good tool selection. For most poured concrete projects, a power concrete mixer is a valuable tool. If the project requires more than one cubic yard of concrete (see chart, page 17), having the concrete delivered in ready-mix form will save a lot of time and back strain, while ensuring a uniform mixing consistency for the entire project. For very small projects and for mixing mortar, a mortar box and masonry hoe can be used effectively.

Use the appropriate alignment and measuring tools to make the layout process easier and more accurate.

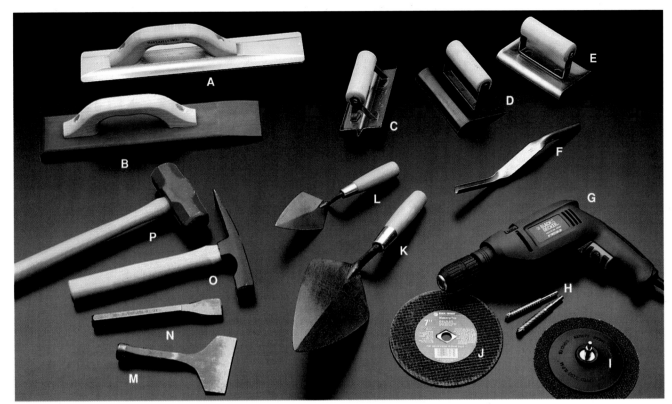

Specialty masonry tools for handling concrete, mortar, and masonry units include: a magnesium float (A) for setting exposed aggregate or smoothing concrete to a hard, glossy finish; a wood float (B) for smoothing most exterior masonry surfaces; a groover (C) for cutting control joints into concrete; a stair edger (D) for creating smooth, even noses on steps; an edger (E) for rounding off edges in concrete at form locations; a jointer (F) for smoothing mortar joints; a power drill (G) with masonry bits (H) and a masonry grinding disc (I); a masonry-cutting blade for a circular saw (J); a masonry trowel (K) for building with brick or block; a pointing trowel (L) for repairing masonry; a brickset (M) for cutting brick and block; a cold chisel (N) for chipping out and breaking apart masonry; a bricklayer's hammer (O) with a claw for cutting masonry units; a maul (P) for driving form stakes and for use with a cold chisel and brickset.

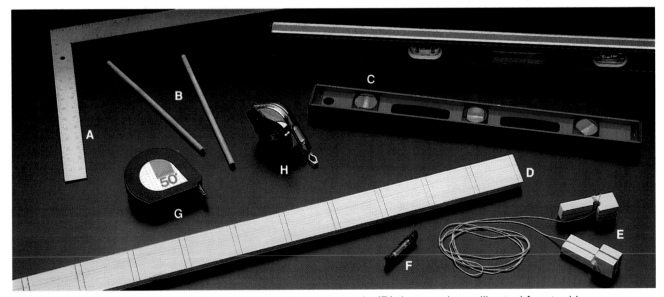

Alignment and measuring tools for masonry projects include: a carpenter's framing square (A) for setting project outlines; ⅜" dowels (B) for use as spacers between dry-laid masonry units; levels (C) for setting forms and for use when stacking masonry units; a story pole (D) that can be calibrated for stacking masonry units; line blocks and mason's string (E) for stacking brick and block; a line level (F) for making layouts and setting slope; a tape measure (G); and a chalkline (H) for marking layout lines on footing or slabs.

Safety & Disposal

Concrete products must be handled with care and disposed of properly. Concrete and mortar mix contains silica, which is a hazardous substance in large quantities, so always wear a particle mask when handling or mixing dry mix, and read and follow the manufacturer's safety precautions. Because these products will irritate your skin, always wear gloves when handling masonry products.

Concrete products are heavy, and lifting and moving them is hard work. For added protection, and when lifting, always wear a lifting belt and take care to use safe lifting techniques.

Most waste-removal companies will not accept masonry waste in regular curbside pickup. You will need to make special pickup arrangements—call your waste-disposal company for information. Where possible, use old masonry waste as clean fill for new projects.

Wear protective equipment, including a particle mask, eyewear, and gloves when mixing masonry products. Concrete products can be health hazards, and they will irritate skin upon contact. Also wear a mask to protect yourself from dust when cutting concrete, brick, or block.

Tips for Working Safely

Wear a lifting belt to help prevent lower back strain when stacking brick and block, and when hand-mixing concrete products. Always lift with your legs, not your back, and keep the items being lifted as close to your body as you can.

Keep the job site clean and well organized by creating a storage area for tools and by removing dirt and debris from the worksite immediately. NOTE: Always use a GFCI extension cord when working with power tools outdoors.

Tips for Working at Heights

Anchor your ladder at the top and bottom, using rope and a screw eye at the top, and stakes at the bottom. Do not carry heavy items up ladders—use a rope and pulley.

Use rented scaffolding for projects that require extended time working at heights, like tuckpointing a chimney. Scaffolding provides a much safer working platform, and is easier on the legs than a ladder. Get safety and operating instructions from the rental store if you have never used scaffolding before. Always make sure the feet of the scaffolding are level and secure (inset).

Tips for Disposal

Save old broken-up concrete for use as clean fill in building and landscaping projects.

Reuse sod by removing it carefully from the project area with a sod cutter (available for rent at most rental stores) or a square-end spade. Roll up the sod carefully, and lay it as soon as possible in another area of your yard.

Poured concrete can be shaped and finished to create a wide variety of surfaces and structures around your home. In the photo above, the steps and the walkway blend together gracefully because the concrete construction allows them to be built essentially as one single unit.

Building with Concrete

Poured concrete is one of the most versatile, durable building materials available. You can use it to make just about any type of outdoor building project. Concrete also costs less than many other building materials, like pressure-treated lumber or brick pavers. Concrete is plain in appearance, but a decorative finish, like exposed aggregate, can enhance its appeal.

Handling concrete can be hard work, but it does not need to be difficult. Timing and preparation are the most important factors. Poured concrete hardens to whatever shape it is in when it dries, whether or not you have finished working on it. Make sure to do thorough site preparation so when you start pouring you can focus your energy and time on placing and smoothing out the fresh concrete—not on staking down forms.

Stick to smaller-scale projects until you have a considerable amount of experience working with concrete. Larger projects, like driveways, de-

mand more expertise and quicker work. But by breaking large projects into several smaller projects, you can combine the simplicity of small projects with the overall scale of larger jobs (see *Exposed-aggregate Patio,* pages 40 to 45).

Get a helper whenever you work with concrete. For best results, pour concrete only when the temperature is between 50° and 80°F.

This sections shows:
- Concrete Basics (pages 16 to 19)
- Preparing the Project Site (pages 20 to 25)
- Placing Concrete (pages 26 to 29)
- Finishing & Curing Concrete (pages 30 to 31)
- Building with Concrete:
  A Step-by-Step Overview (pages 32 to 33)
- Concrete Building Projects
  Walkways (pages 34 to 39)
  Exposed-aggregate Patio (pages 40 to 45)
  Concrete Steps (pages 60 to 65)
  Footing (pages 66 to 69)

Common Concrete Projects

A concrete walkway in a backyard or at a garage or side entrance is a good starter project. The techniques, even for angled walkways like the one shown above, are basic. Because no frost footing is required in most cases, and because walkways can be built to follow gradual slopes, little site preparation is needed. See Walkways, pages 34 to 39.

A patio can be built and finished to blend well with the surrounding elements of just about any yard and house. Using permanent forms between sections allows the entire project to be treated as a series of smaller projects. See Exposed-aggregate Patio, pages 40 to 45.

Concrete steps are long-lasting structures that resist damage from traffic and shoveling. When finished with a non-skid surface, like the broomed surface above, concrete steps are both safe and sturdy. See Concrete Steps, pages 60 to 65.

A poured concrete footing creates a sturdy base for poured concrete projects, as well as for projects built from brick or block. Footings requirements are defined in your local Building Codes. See Footings, pages 66 to 69.

Too dry

Too wet

Correct

Concrete Basics

If you are mixing concrete on-site, you have two options: you can purchase the ingredients separately and blend them together to achieve a mixture suited for your project. Or, you can purchase bags of pre-mixed concrete, and simply add water. For most projects around the home, buying pre-mixed products is a better choice. Premixed concrete yields uniform results, and it is fast and easy to use.

For smaller projects, a wheelbarrow or mortar box is an adequate mixing container. But for larger projects, consider renting or buying a power mixer, or have the concrete delivered by a ready-mix company.

A good mixture of concrete is crucial to any successful concrete project. Properly mixed concrete is damp enough to form in your hand when you squeeze, but not so damp that it loses its shape quickly. If the mixture is too dry, the aggregate will be too hard to work, and will not smooth out easily for an even, finished appearance. A wet mixture will slide off the trowel, and may cause cracking and other defects in the finished surface.

Components of Concrete

The basic ingredients of concrete are the same, whether the concrete is mixed from scratch, purchased premixed, or delivered by a ready-mix company. Portland cement is the bonding agent. It contains crushed lime and cement, and other bonding minerals. Sand and a combination of aggregates add volume and strength to the mix. Water is used to activate the cement. It evaporates to allow the concrete to dry into a solid mass. By varying the ratios of the ingredients, professionals can create concrete with special properties that are suited for specific situations.

Premixed concrete products contain all the components of concrete. Just add water, mix, and you are ready to pour. Usually sold in 60-lb. bags that yield approximately ½ cu. ft., these products are available in different forms with specific properties for specific applications. *General-purpose concrete mix* is usually the most inexpensive product, and is suitable for most do-it-yourself, poured concrete projects. *Fiber-reinforced concrete mix* contains strands of fiberglass that increase strength. For some applications, you can use fiber-reinforced concrete instead of general-purpose concrete to eliminate the need for metal reinforcement. *High-early premixed concrete* contains accelerating agents that cause it to set quickly—a desirable property if you are pouring when temperatures are cold. *Sand mix* contains no mixed aggregate, and is used in repairs and in casting projects where larger aggregate is not desirable.

Tips for Estimating Concrete

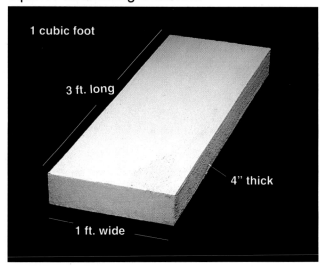

1 cubic foot
3 ft. long
4" thick
1 ft. wide

Concrete Coverage		
Volume	Thickness	Surface Coverage
1 cu. yd.	2"	160 sq. ft.
1 cu. yd.	3"	110 sq. ft.
1 cu. yd.	4"	80 sq. ft
1 cu. yd.	5"	65 sq. ft.
1 cu. yd.	6"	55 sq. ft.
1 cu. yd.	8"	40 sq. ft.

Measure the width and length of the project in feet, then multiply the dimensions to get the square footage. Measure the thickness in feet (4" thick equals ⅓ ft.), then multiply the square footage times the thickness to get the cubic footage. For example, 1 ft. × 3 ft. × ⅓ ft. = 1 cu. ft. Twenty-seven cubic feet equals one cubic yard.

Coverage rates for poured concrete are determined by the thickness of the slab you pour. The same volume of concrete will yield less surface area if the thickness of the slab is increased. The chart above shows the relationship between slab thickness, surface area, and volume.

How to Mix Concrete by Hand

1 Empty contents of premixed concrete bags into a mortar box, wheelbarrow, or another large container. Form a hollow in the pile of dry mix, then pour water into the hollow. Start with 1 gallon of clean tap-water per 60-lb. bag.

2 Work with a hoe, continuing to add water until a good consistency is achieved (page 16). Clear out any dry pockets from the corners. Do not overwork the mix. Also, keep track of how much water you use in the first batch so you will have a reliable recipe for subsequent batches.

How to Mix Concrete with a Power Mixer

1 Fill a bucket with 1 gallon of water for each 60-lb. bag of concrete you will use in the batch (for most power mixers, 3 bags is workable). Pour in half the water. Before you start power-mixing, review the operating instructions carefully.

2 Add all of the dry concrete mix in the batch, then the rest of the water. Mix for a minute. Pour in water as needed until the proper consistency is achieved (page 16), and mix for 3 minutes. Pivot the mixing drum to empty the concrete into a wheelbarrow. Rinse out the drum immediately.

Have ready-mix concrete delivered for large projects. Prepare the site and build the forms yourself, and try to have helpers on hand to help you place and tool the concrete when it arrives.

Ordering Ready-mix Concrete

For large concrete jobs (1 cubic yard or more), have ready-mix concrete delivered to your site. Although it is more expensive, it saves time. Seek referrals, and check your telephone directory under "Concrete" for ready-mix sources.

Tips for preparing for concrete delivery:
• Fully prepare the building site (pages 20 to 25).

• Discuss your project with the experts at the ready-mix company. They will help you decide how much and what type of concrete you need. To help you determine your quantity needs, see the chart on page 17.

• Call the supplier the day before the scheduled pour to confirm the quantity and delivery time.

• Read the receipt you get from the driver. It will tell you at what time the concrete was mixed. Before you accept the concrete, make sure no more than 90 minutes has elapsed between the time it was mixed and the time it was delivered.

Prepare a clear delivery path to the project site, so when the truck rolls up you are ready to pour. Lay planks over the forms and subbase to make a road-way for the wheelbarrows or concrete hoppers. If you have an asphalt driveway, or a concrete driveway that is cracking, have the truck park on the street to prevent driveway damage.

Preparing the Project Site

The basic steps in preparing a project site are: lay out a design using stakes and strings to outline the project; clear the project area and strip off sod; excavate the site to allow for a subbase or a footing, if needed, and the concrete; pour footings or lay a subbase for drainage and stability; build and install wooden forms with reinforcement in place.

Site preparation depends on the type of project and the condition of the site. Plan on using a subbase of compactible gravel. Some projects require footings that extend past the frostline, while others, like sidewalks, do not. Adding metal reinforcement to the project is not always required, but is a good idea. Check with your local building department for guidelines.

Projects built on heavily sloped sites normally require grading of the soil prior to pouring concrete. If your yard slope is more than 1" per ft., you may need to add or remove soil to create a level building surface; contact a landscape engineer or a building inspector for advice on how to modify your yard to accommodate your masonry project.

SAFETY TIP: Beware of buried electric and gas lines when digging. Contact your local public utility company before you start digging.

Everything You Need:

Tools: rope, hand maul, tape measure, mason's string, line level, spade, sod cutter, wheelbarrow, shovel, tamper.

Materials: 2 × 4 lumber, 3" screws, compactible gravel.

Tips for Preparing the Project Site

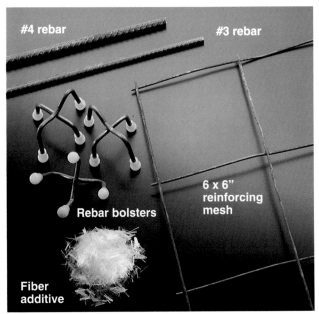

Measure the slope of the building site to determine if you need to do grading work before you start your project. First, drive stakes at each end of the project area. Attach a mason's string between the stakes and use a line level to set it at level. At each stake, measure from the string to the ground. The difference in the distances (in inches), when divided by the distance between stakes (in ft.) will give you the slope (in inches per foot). If the slope is greater than 1" per foot, you likely will need to regrade the building site.

Reinforcement materials: *Metal rebar,* available in sizes ranging from #2 (⅛" diameter) to #5 (⅝" diameter) is used to reinforce narrow concrete slabs, like sidewalks, and in masonry walls. For most projects, #3 rebar (⅜" diameter) is suitable. *Wire mesh* (sometimes called re-mesh) is most common in 6 × 6" grids. It is usually used for broad surfaces, like patios. Rebar and wire mesh are suspended off the subbase by *bolsters. Fiber additive* is mixed into concrete to strengthen small projects that receive little traffic.

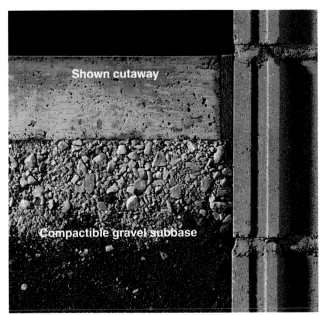

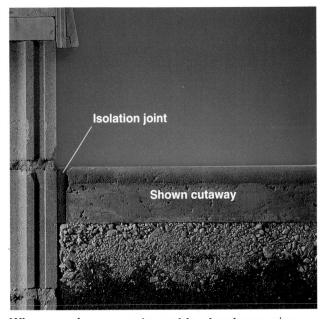

Add a compactible gravel subbase to provide a level, stable foundation for the concrete. The compactible gravel also improves drainage—an important consideration if you are building on soil that is high in clay content. For most building projects, pour a layer of compactible gravel about 5" thick, and use a tamper to compress it to 4" (photo, previous page).

When pouring concrete next to structures, glue a ½"-thick asphalt-impregnated fibrous board, called an isolation board, to the adjoining structure to keep the concrete from bonding with the structure. The board creates an isolation joint, allowing the two structures to move independently, minimizing the risk of cracking.

1 Lay out a rough project outline with a rope or hose. Use a carpenter's square to set perpendicular lines. To create the actual layout, begin by driving wood stakes near each corner of the rough layout. The goal is to arrange the stakes so they are outside the actual project area, but are in alignment with the borders of the project (if you extended the project outlines 1 ft. in each direction, the stakes would be at each of the eight endpoints). NOTE: Projects built next to permanent structures should use the structure to define one project side.

2 Connect the stakes with mason's strings. The strings should follow the actual project outlines. To make sure the strings are square, use the 3-4-5 triangle method: measure and mark points 3 ft. out from one corner along one string, and 4 ft. out along the intersecting string at the corner. Measure between the points, and adjust the positions of the strings until the distance between the points is exactly 5 ft.

3 Reset the stakes, if necessary, to conform to the positions of the squared strings. Check all corners with the 3-4-5 method, and adjust until the entire project area is exactly square. This can be a lengthy process with plenty of trial and error, but it is very important to the success of the project, especially if you plan to build on the concrete surface.

4 Attach a line level to one of the mason's strings to use as a reference. Adjust the string up or down as necessary until it is level. Adjust the other strings until they are level, making sure that intersecting strings contact one another (this ensures that they are all at the same height relative to ground level).

5 Most concrete surfaces should have a slight slope to direct water runoff, especially if they are near your house. To create a slope, shift the level mason's strings on opposite sides of the project downward on their stakes (the lower end should be farther away from the house). To create a standard slope of ⅛" per foot, multiply the distance between the stakes on one side (in ft.) by ⅛. For example, if the stakes are 10 ft. apart, the result will be ¹⁰⁄₈ (1¼"). Move the strings down that much on the stakes on the low ends.

6 Start excavating the project site by removing the sod. Use a sod cutter if you wish to reuse the sod elsewhere in your yard (lay the sod as soon as possible). Otherwise, use a square-end spade to cut away sod. Strip off the sod at least 6" outside the mason's strings to make room for 2 × 4 forms.

7 Make a story pole as a guide for excavating the site. First, measure down to ground level from the high end of a slope line. Add 7½" to that distance (4" for the subbase material and 3½" for the concrete if you are using 2 × 4 forms). Mark the total distance on the story pole, measuring from one end. Remove soil in the project site with a spade. Use the story pole to make sure the bottom of the site is the same distance from the slope line at all points as you dig.

8 Lay a subbase for the project (unless your project requires a frost footing). Pour a 5"-thick layer of compactible gravel in the project site, and tamp until the gravel is even and compressed to 4" in depth. NOTE: The subbase should extend at least 6" beyond the project outline.

How to Build & Install Wood Forms

1 A form is a frame, usually made from 2 × 4 lumber, laid around a project site to contain and shape freshly poured concrete. Cut the 2 × 4s to create a frame with inside dimensions equal to the total size of the project. Gang-cut same-sized boards to save time.

2 Use the mason's strings that outline the project (pages 22 to 23) as a reference for setting form boards in place. Starting with the longest form board, position the boards so the inside edges are directly below the strings.

3 Cut several 12"-long pieces of 2 × 4 to use as form stakes, and trim one end of each stake to create a sharp point. Drive the stakes at the outside edges of the form boards, spaced at 3-ft. intervals, and directly behind any joints in the form boards.

4 Drive 3" deck screws through the stakes and into one of the form boards. Set a level so it spans the staked form and the opposite form board, and use the level as a guide for staking the second form board so it is level with the first (for large projects, use the mason's strings as the primary guide for setting the height of all form boards).

5 Stake the rest of the form boards to complete the frame. With the tops of all form boards level, drive 3" deck screws at all corner joints to bond the form frame together. Coat the forms with vegetable oil before you pour concrete. TIP: Tack nails or drive screws into the sides of the forms to mark the control joint locations, saving time during the pour.

Variations for Building Forms

⅛" hardboard

Use plywood (top left photo) for building taller forms for projects like concrete steps (pages 60 to 65). Gang-cut plywood form sides, and brace with 2 × 4 arms attached to 2 × 4 cleats at the sides.
Use the earth as a form (bottom left photo) when building footings for poured concrete building proj-

ects (page 67). Use standard 2 × 4 forms for the tops of footings for building with brick or block.
Create curves (above, right) with ⅛"-thick hardboard attached at the inside corners of a form frame. Drive support stakes behind the curved form.

Tips for Laying Metal Reinforcement

Cut metal rebar with a reciprocating saw that is equipped with a metal-cutting blade (cutting metal rebar with a hacksaw can take 5 to 10 minutes per cut). Use wire cutters to cut wire mesh.

Overlap joints in metal rebar by at least 12", then bind the ends together with heavy-gauge wire. Overlap seams in wire mesh reinforcement by 12".

Stay at least 1" away from forms with the edges or ends of metal reinforcement. Use bolsters or small chunks of concrete to raise metal reinforcement off the subbase, but make sure it is at least 2" below the tops of the forms.

Placing Concrete

Placing concrete involves delivering fresh concrete from the source into forms, then beveling and smoothing the concrete with a variety of masonry tools. Once the surface is smoothed and level, control joints are cut and the edges of the project are rounded off for a finished look. Because placing and tooling directly affect the outward appearance of any concrete building project, it is important that you do careful work.

> ### Everything You Need:
>
> Tools: wheelbarrow, hoe, spade, hammer, trowel, wood float, groover, edger.
>
> Materials: concrete, 2 × 4 lumber, mixing container, water container.

Start pouring concrete at the farthest point from the concrete source, and work your way back.

Tips for Pouring Concrete

Do not overload your wheelbarrow. Experiment with sand or dry mix before mixing concrete to find a comfortable, controllable volume. This also helps you get a feel for how many wheelbarrow loads it will take to complete your project.

Lay planks over the forms to make a ramp for the wheelbarrow. Avoid disturbing the building site by using ramp supports (above). Make sure you have a flat, stable surface between the concrete source and the forms.

How to Place Concrete

1 Load the wheelbarrow with fresh concrete. Make sure you have a clear path from the source to the site. Always load wheelbarrows from the front. Loading wheelbarrows from the sides can cause tipping.

2 Pour concrete in evenly spaced loads (each load is called a "pod"). Start at the end farthest from the concrete source, and pour so the top of the pod is a few inches above the tops of the forms. Do not pour too close to the forms. NOTE: If you are using a ramp, stay clear of the end of the ramp.

3 Continue to place concrete pods next to preceding pods, working away from the first pod. Do not pour more concrete than you can tool at one time. Keep an eye on the concrete to make sure it does not harden before you can start tooling.

4 Distribute concrete evenly in the project area, using a masonry hoe. Work the concrete with a hoe until it is roughly flat, and the surface is slightly above the tops of the forms.

(continued next page)

5 Work the blade of a spade between the inside edges of the form and the concrete to remove trapped air bubbles that can weaken concrete.

6 Rap the forms with a hammer or the blade of the shovel to help settle the concrete. This also draws finer aggregates in the concrete against the forms, creating a smoother surface on the sides. (This is especially important when building steps.)

7 Use a screed board—a straight piece of 2 × 4 long enough to rest on opposite forms—to strike off the excess concrete. Move the screed board in a sawing motion from left to right, and keep the screed flat as you work. If screeding leaves any valleys in the surface, add fresh concrete in the low areas and screed to level.

8 Precut control joints (page 24, step 5) at pre-marked locations with a mason's trowel, using a straightedge as a guide. Control joints are cut into the concrete surface (see step 10) to direct cracking in a direction that does not cause structural damage. Space the control joints evenly for a more attractive appearance.

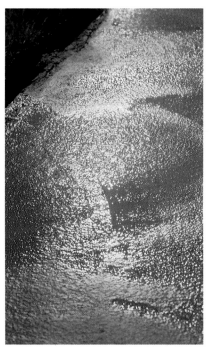

9 Use a wood float to smooth the surface. Float lightly, using an arcing motion, until the entire surface is smooth. Do not overfloat. NOTE: When tooling concrete, tip the lead edge of the tool upward slightly to prevent gouging.

TIP: Floating concrete causes puddles of water, called "bleed water," to form on the surface. Once bleed water forms, finish up any tooling as quickly as possible.

10 Allow the concrete to dry until the bleed water disappears. Then draw a groover across the precut control joints (step 8), using a straightedge as a guide. You may need to make several passes to create a smooth control joint.

11 Shape concrete with an edging tool between the forms and the concrete to create a smooth, finished appearance. You may need to make multiple passes. Use a wood float to smooth out any marks left by the edger or groover. Apply decorative finishes, if desired, then let the concrete cure (pages 30 and 31).

Finishing option: exposed aggregate. Sometimes called "seeding" concrete, applying decorative aggregate to the fresh concrete surface creates an attractive effect with many design options. The photo above shows what a few varieties of aggregate look like when they are used for an exposed-aggregate surface.

Finishing option: broomed finish. Create by dragging a stiff-bristled garage broom across a concrete surface immediately after tooling the concrete. Do not overwork the concrete.

Finishing & Curing Concrete

Finishing and curing concrete are critical last steps in any concrete project. Proper curing ensures that the concrete reaches its maximum strength and remains free of surface defects that can ruin the appearance. There are many theories on the best way to cure concrete, but for residential projects, simply covering the concrete with plastic is a simple and effective method.

Applying a decorative finish dresses up the plain appearance of concrete surfaces. Exposed-aggregate (or "seeded") finishes are common on walkways, patios, and other surfaces. Brooming is a good option for any surface that receives high traffic volume. Check around your neighborhood for other ideas for creative concrete finishes.

Everything You Need:
Tools: broom, wheelbarrow, shovel, wood float, hand broom, hose, coarse brush.

Materials: sheet plastic, "seeding" aggregate, water.

How to Cure Concrete

Use a plastic covering for curing concrete. Curing concrete keeps the water in the concrete from evaporating too quickly, which helps prevent surface defects. Cure for a week, covered with plastic. Anchor the plastic securely, and overlap and tape any seams.

How to Create an Exposed-aggregate Finish

1 After smoothing off the surface with a screed board (page 28), spread aggregate evenly over the surface with a shovel or by hand. Smaller aggregate (up to 1" in diameter) should be spread in a single-thickness layer; for larger seeding stones, try to maintain a separation between stones that is roughly equal to the size of one stone. NOTE: Always wash the aggregate thoroughly before seeding.

2 Float the seeded surface with a magnesium float until a thin layer of concrete rises up to completely cover the stones. Do not overfloat. TIP: Cover the surface with plastic to keep the concrete from hardening too quickly if you are seeding a large area.

3 As soon as the bleed water disappears, cut control joints and tool the edges (page 29). Let the concrete set for 30 to 60 minutes, then mist a section of the surface and scrub with a brush to remove the concrete covering the aggregate. If any stones dislodge, reset them and try again later. When you can scrub without dislodging stones, mist and scrub the entire surface to expose the aggregate. Rinse clean. Do not let the concrete dry too long, or it will be difficult to scrub off.

4 After the concrete has cured for one week (previous page), remove the covering and rinse the surface with a hose. If any cement residue remains, try scrubbing it clean. If scrubbing is ineffective, wash the surface with a muriatic acid solution, then rinse immediately and thoroughly with water. OPTION: After three weeks, apply exposed-aggregate sealer (pages 70 to 71).

![Image]()

Building with Concrete: A Step-by-Step Overview

The project shown on these pages provides a basic overview of the sequence of techniques and processes used in building with concrete. The project features the pouring of a simple concrete slab that will serve as a base for a decorative-block screen (see pages 92 to 93 for construction of the screen). Refer to the cited pages for more information on the steps shown.

Because the project shown serves as a base for concrete block, we dry-laid the block on the project site to determine the size of the slab (page 92).

1 Create a plan for the project (pages 8 to 9 and 16 to 19), then use stakes and mason's strings to outline the project. In the photo above, the 3-4-5 method of squaring up a layout is being used (page 22). After the layout is set, remove sod and excavate soil in the project site (page 23).

2 Create a subbase for the project by tamping compactible gravel to a consistent depth. Use a story pole to guide your work (page 23).

3 Install 2 × 4 forms around the project site (pages 24 to 25). Attach isolation boards to any permanent structures that will adjoin the project (page 21). Install any reinforcement required for the project (page 25). NOTE: If your project requires a footing, build it before preparing the subbase.

4 Calculate the volume of concrete needed for the job (see chart, page 17), and mix the concrete or have it delivered to the work site. Place the concrete into the forms, then use a masonry hoe to distribute it evenly. Work around the inside edges of the forms with a shovel and rap the forms with a hammer to release air bubbles (page 28).

5 Level off the concrete so it is even with the tops of the forms, using a screed board, then smooth the surface with a wood float (page 29). Let the concrete dry until all water evaporates from the surface (page 29). If you plan to apply a decorative finish, like exposed aggregate (pages 30 to 31), take care of it before the concrete hardens.

6 Shape the edges of the concrete at the forms, using an edger tool, and form control joints in larger slabs, using a groover (page 29). Cover the concrete with plastic and let it cure for a week.

7 Remove the plastic and the forms, then backfill up to the edges of the concrete with dirt. Apply concrete sealer, if desired (pages 70 to 71).

An angled concrete walkway creates a functional link between two points, while adding a unique design element to your yard.

Concrete Building Projects

Walkways

If your house fills up with muddy footprints after every rainstorm, and a path of worn grass has developed in your yard, the chances are good that you need a walkway. Poured concrete is an ideal material for residential walkways since nothing matches it for permanence and resistance to damage from foot traffic or shoveling.

Always check with your local building department before starting a walkway project. This is especially important if you plan to build a walkway in the front of your house. Most areas closely regulate the construction of public-access sidewalks.

Tips for Planning:

• Design walkways to follow traffic flow, but bear in mind that a walkway can divide a yard. You may prefer to route the walkway closer to the edge of your property to preserve larger areas of your yard for recreational use.

• Build walkways that are the same width as other existing walkways (3 ft.and 4 ft. are common widths; 2 ft. is the recommended minimum width).

• A walkway can follow a gradual slope in a yard, but severe slopes require steps or regrading.

• Keep walkways at least 2 ft. away from trees. Damage from roots and tree trunks is a leading cause of concrete failure.

Tips for Directing Water Runoff on Walkways

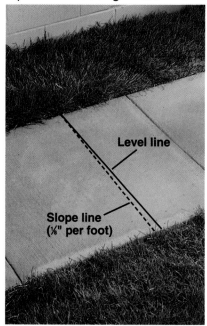

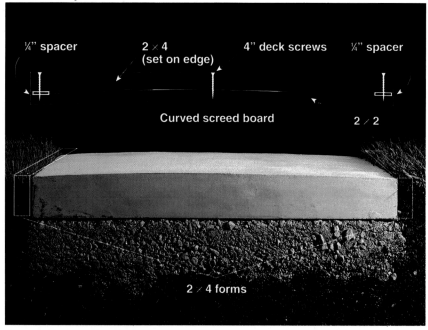

Slope walkways away from the house to prevent water damage to the foundation or basement. Outline the location of the walkway with mason's strings, then lower the outer string to create a slope of ⅛" per ft. (see pages 22 to 23).

Crown the walkway so it is ¼" higher at the center than at the edges. This will prevent water from pooling on the surface. To make the crown, construct a curved screed board by cutting a 2 × 2 and a 2 × 4 long enough to rest on the walkway forms. Butt them together edge to edge and insert a spacer between them at each end. Attach the parts with 4" deck screws driven at the center and the edges. The 2 × 2 will be drawn up at the center, creating a curved edge. Smooth the concrete with the curved edge of the screed board facing down.

Tips for Designing Walkways

Use curves to add visual interest to an otherwise plain walkway or to add contrast to yards with mostly square features. Use the same construction techniques for straight or angled walkways, but use ⅛"-thick hardboard to make the curved forms (page 25).

Look for traffic patterns in your lawn to find the most logical location for a new walkway.

1 Select a rough layout for the walkway based on the design tips shown on page 34. For function and appearance, we chose to include an angled turn in our walkway. Stake out the location and connect the stakes with mason's strings. Set slope lines if needed (pages 22 to 23).

2 Remove sod in the project site and 6" beyond the outlines, then excavate the site with a spade, following the slope lines to maintain consistent depth (page 23).

3 Pour a 5"-thick layer of compactible gravel into the site to create a subbase for the walkway. Tamp the subbase until it compacts to 4" thick and it is even on the surface (page 23).

4 Build and install forms made out of 2 × 4 boards set on edge (page 24). Miter-cut the ends at angled joints. Position them so the inside edges are lined up with the strings, then drive 2 × 4 stakes next to the forms at 3-ft. intervals. Attach the stakes to the forms with 3" deck screws. Use a carpenter's level to make sure the forms are level to one another. Drive a stake at each side of angled form joints.

5 Glue an isolation board (page 21) to steps, the house foundation, or other permanent structures that adjoin the walkway.

OPTION: Reinforce the walkway with #3 steel rebar (page 33). For a 3-ft.-wide walkway, lay two sections of rebar spaced evenly inside the project area. Use bolsters to support the rebar (make sure bolsters are at least 2" below the tops of the forms). Bend rebar to follow any angles or curves, and overlap pieces at angled joints by 12". Mark locations for control joints (to be cut with a groover later) by tacking nails to the outside faces of the forms, spaced roughly at 3-ft. intervals.

(continued next page)

6 Mix, then pour concrete into the project area (pages 26 to 27). Use a masonry hoe to spread it evenly within the forms. After pouring all of the concrete, run a spade along the inside edges of the form, then rap the outside edges of the forms with a hammer to help settle the concrete.

7 Build a curved screed board (page 35) and use it to form a crown when you smooth out the concrete (page 28). NOTE: A helper makes this easier.

8 Smooth the surface with a wood float (page 29). Precut control joints at marked locations (page 28) using a trowel and a straightedge. Let the concrete dry until the bleed water disappears (page 29).

9 Shape the edges of the concrete by running an edger along the forms. Smooth out any marks created by the edger with a float. Lift the leading edges of the edger and float slightly as you work.

10 Cut control joints using a groover and a straightedge as a guide. Use a float to smooth out any tool marks.

11 Create a textured, non-skid surface by drawing a clean, stiff-bristled broom across the surface. Avoid overlapping broom marks. Cover the walkway with plastic and let the concrete cure for one week (pages 30 to 31).

12 Remove the forms, then backfill the space at the sides of the walkway with dirt or sod. Seal the concrete, if desired (pages 70 to 71), according to the manufacturer's directions.

This four-square patio replaced a small, crumbling slab that had become an eyesore. The new patio combines pressure-treated wood with aggregate to create a natural-looking, durable, surface.

Concrete Building Projects
Exposed-aggregate Patio

Enhance your outdoor living space and beautify your yard by adding an attractive concrete patio. Dividing this 10 ft. × 10 ft. area into four even quadrants, separated by permanent forms, created a strong patio. At the same time, it made it possible to complete the project in four easy stages. Each quadrant can be poured, tooled and seeded separately.

Footings are not needed for this project, since we use an isolation joint to separate the patio from the foundation. Do careful preparation work and add a thick subbase of compactible gravel to improve stability and drainage.

NOTE: Portions of this project were used to demonstrate project layout on pages 22 to 25. Refer to those pages before starting your project.

Tips for Planning:

• Design the patio to be roomy: allow at least 20 sq. feet of patio space for each regular user.

• Lay newspapers or blankets in the proposed building area to help you visualize patio shapes and sizes in relation to the rest of your yard and your house.

• Where possible, plan border plantings or other decorative elements to help separate the patio from the yard.

• Incorporate fences, trellises, or other backyard building structures to create an outdoor "room," especially if you have nearby neighbors.

• Create a gentle slope away from your house (⅛" per foot is plenty). Do not exceed more than 1" per foot in slope.

Design Tips for Patios

Maintain plenty of clearance between the top of the patio and the door threshold. The top of the patio should be at least 2" below the house sill or threshold so the concrete has room to expand without causing damage in the event of frost heave:

Integrate landscaping into the patio design. The exposed-aggregate surface on this patio seems to almost flow into the adjoining greenery and rock garden.

Incorporate steps or retaining walls into the patio. Originally, the yard in the photo above sloped steeply away from the house. By constructing a retaining wall and integrating steps into the wall, the designer of this masonry project was able to fill in the slope and build a patio near the threshold level.

Build a "floating" patio away from the house. Building a patio in a remote location is a good way to create a separate, defined area for entertaining or relaxing. Floating patios can be built in just about any shape or size.

How to Build an Exposed-aggregate Patio

1 Prepare the building site by removing any existing building materials, like sidewalks or landing areas. Consider the design tips on the previous page and throughout this book, and select a design for your concrete patio.

2 Lay out the rough position of the patio using a rope or hose, then mark the exact layout with stakes and mason's strings. Use the 3-4-5 method to make sure the project area is square (page 22). Establish a ⅛" per ft. slope away from the house (pages 22 to 23), if the ground is level.

3 Remove sod and excavate the project area to a consistent depth, using mason's strings and a story pole as reference (page 23).

4 Create a subbase for the patio by pouring a 5"-thick layer of compactible gravel, then tamping until it is even and 4" thick.

5 Cut brown pressure-treated 2 × 4 boards to build a permanent form frame that outlines the entire patio site. Lay the boards in place, using the mason's strings as guides. Fasten the ends together with 2½" galvanized deck screws. Temporarily stake the forms at 2-ft. intervals. Set a straight 2 × 4 across the side forms, and set a level on top of the 2 × 4 to check the side forms for level. SAFETY TIP: Wear gloves and a particle mask when cutting pressure-treated lumber.

Half-length form boards

Full-length form board

6 Cut and install pressure-treated 2 × 4s to divide the square into quadrants: cut one piece full length, and attach two one-half length pieces to it with screws driven toenail style. Drive 4" deck screws partway into the forms every 12" at inside locations. The portions that stick out will act as tie rods between the poured concrete and the permanent forms. TIP: Protect the tops of the permanent forms by covering them with masking tape.

7 Cut reinforcing wire mesh to fit inside each quadrant, leaving 1" clearance on all sides (page 25). Mix concrete and pour the quadrants one at a time, starting with the one located farthest from the concrete source (pages 26 to 29). Use a masonry hoe to spread the concrete evenly in the forms.

8 Smooth off the concrete using a straight 2 × 4 as a screed board (page 28). Settle the concrete by sliding a spade along the inside edges of the forms and rapping the outer edges with a hammer.

(continued next page)

How to Build an Exposed-aggregate Patio (continued)

9 After striking off the surface, cover the concrete with a full layer of aggregate "seeds." Use a magnesium float to embed the aggregate completely into the concrete (pages 30 to 31).

10 Tool the edges of the quadrant with an edger, then use a wood float to smooth out any marks left behind. TIP: If you plan to pour more quadrants immediately, cover the seeded concrete with plastic so it does not set up too quickly.

11 Pour the remaining quadrants, repeating steps 7 through 10. Check poured quadrants periodically; if the bleed water has evaporated significantly, uncover the quadrants and proceed to step 12.

12 After the last of the bleed water has evaporated, expose the aggregate by misting the surface with water and scrubbing the surface with a stiff-bristled broom. Remove the protective tape from the forms, then re-cover the quadrants with plastic and let the concrete cure for one week.

13 After the concrete has cured, rinse and scrub the aggregate again to clean off any remaining residue. TIP: Use diluted muriatic acid to wash off stubborn concrete residue. Read manufacturer's instructions for mixing ratios and safety precautions.

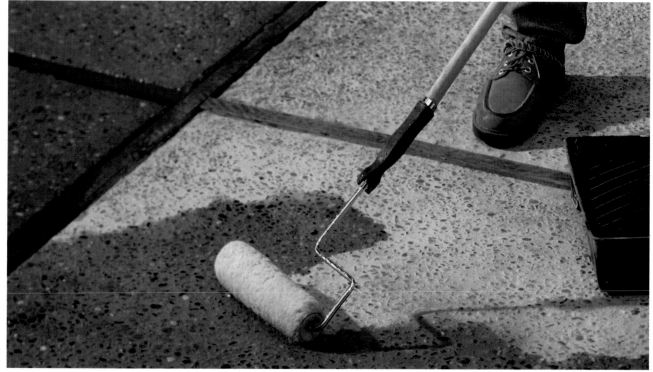

14 Wait three weeks, then seal the patio surface with exposed-aggregate sealer. Reapply sealer periodically as needed, following manufacturer's recommendations.

Patio tile can turn a drab concrete slab into a charming outdoor living area. To create this tiled project, we first poured a new concrete subbase over an existing concrete patio (inset).

New subbase

Old patio

Installing Tiled Patios

If you have ever laid ceramic or vinyl tile inside your house, you already have valuable experience that will help you lay patio tile.

The primary difference between interior and exterior tile is in the thickness of the tiles and the water-absorption rates. The project layout and application techniques are quite similar. With patio tiling projects, however, preparing or creating a suitable subbase for the tiles can become a fairly intensive project. Because any exterior project must stand up to the elements, a sturdy subbase is critical.

Patio tile is most frequently applied over a concrete subbase—either an existing concrete patio, or a new concrete slab. A third option, which we show you on the following pages, is to pour a new tile subbase over an existing concrete patio. This option involves far less work and expense than removing an old patio and pouring a new slab. And it ensures that your new tiled patio will not develop the same problems that

may be present in the existing concrete surface. See the photographs at the top of page 48 to help you determine the best method for preparing an existing concrete patio for tile.

If you do not have a concrete slab in the project area already, you will need to pour one before you tile (see pages 44 to 49).

The patio tiling project shown here is divided into two separate projects: pouring a new subbase, and installing patio tile. If your existing patio is in good condition, you do not need to pour a new subbase.

When selecting tiles for your patio, make sure the product you purchase is exterior tile, which is designed to better withstand freezing and thawing than interior tile. Try to select colors and textures that match or complement other parts of your house and yard. If your project requires extensive tile cutting, arrange to have the tiles cut to size at the supply center.

Exterior tile products for patios are denser and thicker than interior tile. Common types include shell-stone tile, ceramic patio tile, and quarry tile. The most common size is 12" × 12", but you also can purchase precut designer tiles that are assembled into elaborate patterns and designs.

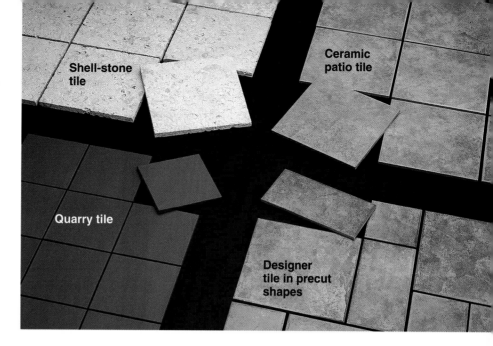

Shell-stone tile

Ceramic patio tile

Quarry tile

Designer tile in precut shapes

Tools for working with exterior tile include: a wet saw for cutting large amounts of tile (usually a rental item); a square-notched trowel for spreading tile adhesive; a grout float for spreading grout into joints between tiles; a sponge for wiping up excess grout; tile nippers for making curved or angled cuts in tiles; tile spacers to set standard joints between tiles; and a rubber mallet for setting tiles into adhesive.

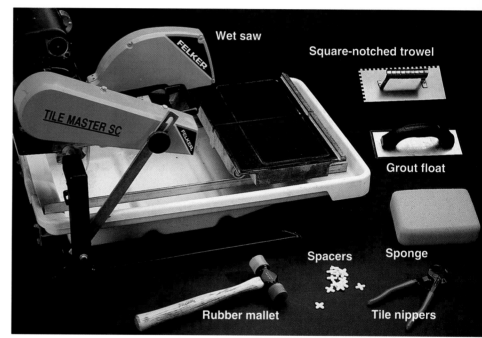

Wet saw

Square-notched trowel

Grout float

Sponge

Spacers

Rubber mallet

Tile nippers

Materials for installing patio tile include: exterior tile grout (tinted or untinted); acrylic grout sealer; latex caulk for filling tile joints over control joints; caulk backer rod to keep grout out of control joints during grout application; latex-fortified grout additive; tile sealer; concrete floor-mix for building a tile subbase; tile adhesive (dry-set mortar); and a mortar bag for filling joints with grout (optional).

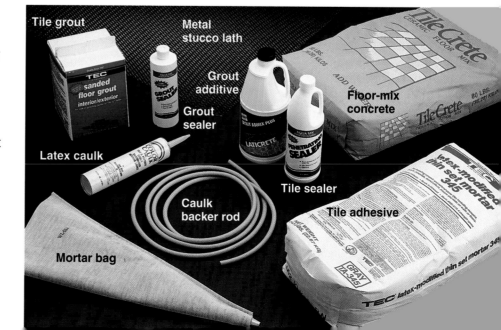

Tile grout

Metal stucco lath

Grout additive

Grout sealer

Floor-mix concrete

Latex caulk

Tile sealer

Caulk backer rod

Tile adhesive

Mortar bag

Tips for Evaluating Concrete Surfaces

A good surface is free from any major cracks or badly flaking concrete (called spalling). You can apply patio tile directly over a concrete surface that is in good condition if it has control joints (see below).

A fair surface may exhibit minor cracking and spalling, but has no major cracks or badly deteriorated spots. Install a new concrete subbase over a surface in fair condition before laying patio tile.

A poor surface contains deep or large cracks, broken, sunken or heaved concrete, or extensive spalling. If you have this kind of surface, remove the concrete completely and replace it with a new concrete slab before you lay patio tile.

Tips for Cutting Control Joints in a Concrete Patio

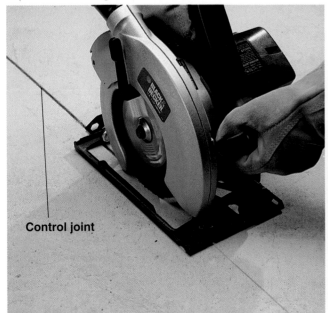

Control joint

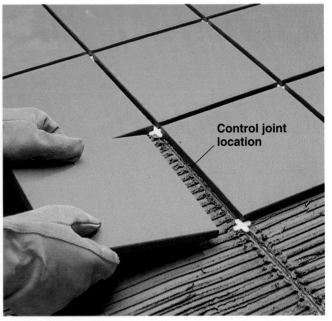

Control joint location

Cut new control joints into existing concrete patios that are in good condition (see above) but do not have enough control joints—control joints manage cracking in concrete by directing cracks in directions that do not weaken the concrete or detract from the appearance. They should be cut every 5 or 6 ft. in a patio. Use a circular saw with a masonry blade set to ⅜" depth to cut control joints. If possible, plan the control joints so they will be below tile joints once the tile layout is established.

How to Install a Subbase for Patio Tile

Pour a layer of floor-mix concrete over an old concrete patio to create a level tile subbase free from cracks that can cause new tile to fail.

Pouring a concrete subbase over an old concrete slab is similar to other concrete projects, but with a few important differences. As with most projects, you will need to: install wooden forms around the perimeter of the project area, smooth the concrete surface, and cure the concrete.

Elements unique to this project include: the use of floor-mix concrete (a dry mix designed for pouring a subbase), the need for a bond-breaker (we laid 30# building paper over the old surface to keep the new surface from adhering to it), and special reinforcement requirements (we used ⅜"-thick metal stucco lath).

Everything You Need:

Tools: basic hand tools, shovel, maul, straight-edge, masonry hoe, mortar box, tamper, magnesium float, concrete edger, utility knife, trowel.

Materials: 30# building paper, sheet plastic, 2 × 4 lumber, stucco lath, tile subbase, floor mix, roofing cement.

Form braces

2 × 4 form boards

1 Dig a trench around the old patio to create room for 2 × 4 forms. Make the trench at least 6" wide, and no more than 4" deep. Clean all dirt and debris from the exposed sides of the patio. Cut 2 × 4s the same length as the sides of the patio, then measure the side-to-side distance, add 3" to the length of the side forms, and cut 2 × 4s for the front—and for the back if the patio is not next to the house. Lay 2 × 4s around the patio, on edge, and join the ends with 3" deck screws. Cut several 12"-long wood stakes and drive them next to the forms, at 2-ft. intervals.

2 × 2 spacer

Stucco lath (used as a spacer)

2 Adjust form height: set a stucco lath spacer on the surface, then set a 2 × 2 spacer on top of the lath (their combined thickness equals the thickness of the subbase). Adjust the tops of the form boards so they are level with the 2 × 2 spacer, and screw the braces to the forms.

Building paper "bond-breaker"

3 Remove the spacers, then lay strips of 30# building paper over the old patio surface, overlapping seams by 6", to create a "bond-breaker" for the new surface (this prevents the new subbase from bonding directly to the old concrete). Crease the building paper at the edges and corners, making sure the paper extends past the tops of the forms. Make a small cut in the paper at each corner so the paper can be folded over more easily.

(continued next page)

4 Lay strips of stucco lath over the building-paper bond-breaker, overlapping seams by 1". Keep the lath 1" away from forms and the wall. Use aviator snips to cut the stucco lath (wear heavy gloves when handling metal).

5 Install temporary 2 × 2 forms to divide the project into working sections and provide rests for the "screed board" used to level and smooth the fresh concrete. Make the sections narrow enough that you can reach across the entire section (3-ft. to 4-ft. sections are comfortable for most people). Screw the ends of the 2 × 2s to the form boards so the tops are level.

6 Mix dry floor-mix concrete with water in a mortar box, blending with a masonry hoe, according to the manufacturer's directions. The mixture should be very dry when prepared (inset) so it can be pressed down into the voids in the stucco lath with a tamper.

7 Fill one working section with floor-mix concrete, up to the tops of the forms. Tamp the concrete thoroughly with a lightweight tamper to help force it into the voids in the lath and into corners. The lightweight tamper shown above is made from a 12" × 12" piece of ¾" plywood, with a 2 × 4 handle attached.

8 Level off the surface of the concrete by dragging a straight 2 x 4 "screed board" across the top, with the ends riding on the forms. Move the 2 x 4 in a sawing motion as you progress. Called "screeding" or "striking-off," this process creates a level surface and fills any voids in the concrete. If any voids or hollows remain, add more concrete and smooth it off.

9 Use a wood float to smooth out the surface of the concrete. Applying very light pressure, move the float back and forth in an arching motion, tipping the lead edge up slightly to avoid gouging the surface. Float the surface a second time, using a magnesium float, to create a hard, smooth surface.

10 Pour and smooth out the next working section, repeating steps 7 to 9. After floating this section, remove the 2 x 2 temporary form between the two sections. Fill the void left behind with fresh concrete. Float the fresh concrete with the magnesium float until it is smooth, level, and it blends into the working section on each side. Pour and finish the remaining working sections one at a time, using the same techniques.

(continued next page)

How to Install a Subbase for Patio Tile (continued)

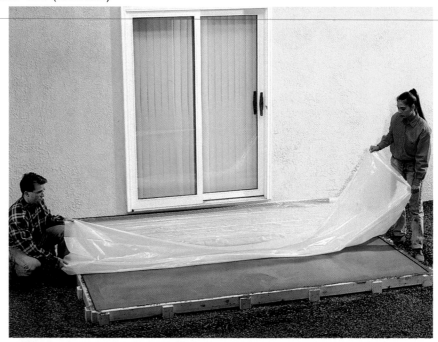

11 Let the concrete dry until pressing the surface with your finger does not leave a mark. Cut contours around all edges of the subbase with a concrete edger. Tip the lead edge of the edger up slightly to avoid gouging the surface. Smooth out any marks left by the edger, using a float.

12 Cover the concrete subbase with sheets of plastic, and cure for at least three days (see manufacturer's directions for recommended curing time). Weight down the edges of the sheeting. After curing is compete, remove the plastic and disassemble and remove the forms.

13 Trim off the building paper around the sides of the patio, using a utility knife. Apply plastic roof cement to the sides of the patio with a putty knife or trowel, to fill and seal the seam between the old surface and the new surface. To provide drainage for moisture between layers, do not seal the lowest side of the patio. After the roof cement dries, shovel dirt or groundcover back into the trench around the patio.

Installing Tiled Patios

How to Lay Patio Tile

With any tiling project, the most important part of the job is creating and marking the layout lines and pattern for the tile. The best way to do this is to perform a dry run using the tiles you will install. Try to find a layout that requires the least possible amount of cutting.

Once you have established an efficient layout plan, carefully mark square reference lines in the project area. Creating a professional-looking tiled patio requires that you follow the lines and use sound installation techniques.

Some patio tile is fashioned with small ridges on the edges that automatically establish the spacing between tiles. But more often, you will need to insert plastic spacers between tiles as you work. The spacers should be removed before the tile adhesive dries.

Tiled patios are vulnerable to cracking. Make sure the tile subbase has sufficient control joints to keep cracking in check (page 48). Install tiles so the control joints are covered by tile joints. Fill the tile joints over the control joints with flexible latex caulk, not grout.

Everything You Need:

Tools: carpenter's square, straightedge, tape measure, chalk line, tile cutter, tile nippers, tile spacers, trowel, rubber mallet, grout float, grout sponge, caulk gun.

Materials: buckets, paint brush and roller, plastic sheeting, paper towels, dryset mortar, tile, backer rod, grout, grout additive, grout chalk, grout sealer, tile sealer.

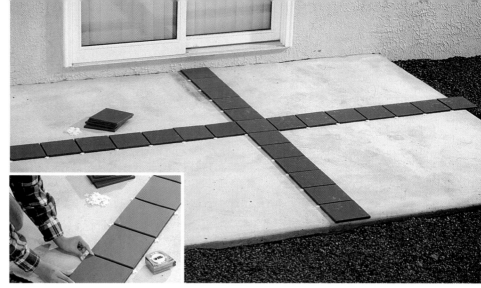

1 Without applying adhesive, set rows of tile onto the surface so they run in each direction, intersecting in the center of the patio. Slip tile spacers between tiles to represent joints (inset). This "dry lay" helps you establish and mark an attractive, efficient layout for the tile.

2 Adjust the dry lay to create a layout that minimizes tile cutting. Shift the rows of tiles and spacers until the overhang is equal at each end, and any cut portions are less than 2" wide. NOTE: Tiles should be ¼" to ½" away from the house.

Snap chalk line for reference

3 Once the layout is set, mark layout lines onto the tiling surface. Mark the surface at the joint between the third and fourth row out from the house, then measure the distance and mark it at several more points along the project area. Snap a chalk line to connect the marks.

(continued next page)

4 Use a carpenter's square and a long, straight board to mark end points for a second reference line perpendicular to the first. Mark the points next to the dry-laid tile so the line falls on a joint location. Remove tools and tiles, and snap a chalkline that connects the points.

5 Lay tiles in one quadrant at a time, beginning with a section next to the house. Start by mixing a batch of dry-set mortar in a bucket, according to the manufacturer's directions. Spread mortar evenly along both legs of one quadrant, using a square-notched trowel. Apply enough mortar for four tiles along each leg.

6 Use the edge of the trowel to create furrows in the mortar. Make sure you have applied enough mortar to completely cover the area under the tiles, without covering up the reference lines.

7 Set the first tile in the corner of the quadrant where the lines intersect, pressing down lightly and twisting slightly from side to side. Adjust the tile until it is exactly aligned with both reference lines.

8 Rap the tile gently with a rubber mallet to set it into the mortar. Rap evenly across the entire surface area, being careful not to break the tile or completely displace the mortar beneath the tile. NOTE: Once you start to fill in the "field" area of the quadrant, it is faster to place several tiles at once, then set them all with the mallet at one time.

9 Set plastic spacers at the corner of the tile that faces the working quadrant. NOTE: Plastic spacers are only temporary: be sure to remove them before the mortar hardens—usually within one hour.

10 Position the next tile into the mortar bed along one arm of the quadrant, making sure the tiles fit neatly against the spacers. Rap the tile with the mallet to set it into the mortar, then position and set the next tile on the other leg of the quadrant. Make certain the tiles align with the reference lines.

11 Fill out the rest of the tile in the mortared area of the quadrant, using the spacers to maintain uniform joints between tiles. Wipe off any excess mortar before it dries.

(continued next page)

12 Apply a furrowed layer of mortar to the field area: do not cover more area than you can tile in 15 to 20 minutes. TIP: Start with smaller sections, then increase the size as you get a better idea of your working pace.

13 Set tiles into the field area of the first quadrant, saving any cut tiles for last. Rent a wet saw from your local rental store for cutting tiles, or have them cut to size. To save time, have all tiles cut before you start laying tiles.

14 Apply mortar and fill in tiles in the next quadrant against the house, using the same techniques used for the first quadrant. Carefully remove plastic spacers with a screwdriver as you finish each quadrant—do not leave spacers in mortar for more than one hour.

15 Fill in the remaining quadrants. TIP: Use a straightedge to check the tile joints occasionally. If you find that any of the joint lines are out of alignment, compensate for the misalignment over several rows of tiles.

16 After all the tiles for the patio are set, check to make sure all spacers are removed and any excess mortar has been cleaned from the tile surfaces. Cover the project area with plastic for three days to allow the mortar to cure properly.

Caulking backer rod

17 After three days, remove the plastic and prepare the tile for "grouting" (the process of filling the joints between tiles with grout). Create expansion joints on the tiled surface by inserting strips of ¼"-diameter caulking backer rod into the joints between quadrants, and over any control joints (page 48), to keep grout out of these joints.

(continued next page)

18 Mix a batch of tile grout to the recommended consistency. TIP: Add latex-fortified grout additive so excess grout is easier to remove. Starting in a corner and working out, pour a layer of grout onto an area of the surface that is 25 square feet or less in size. Spread out the grout with a rubber grout float so it completely fills the joints between tiles.

19 Use the grout float to scrape off excess grout from the surface of the tile. Scrape diagonally across the joints, holding the float in a near-vertical position. Patio tile will absorb grout quickly and permanently, so it is important to remove all excess grout from the surface before it sets.

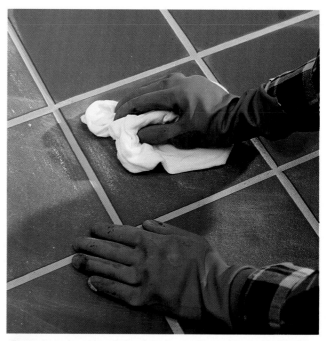

20 Use a damp sponge to wipe the grout film from the surface of the tile. Rinse the sponge out frequently with cool water, and be careful not to press down so hard around joints that you disturb the grout. Wash grout off of the entire surface.

21 Let the grout dry for about four hours, then test by poking with a nail to make sure it is completely hardened. Use a cloth to buff the surface of the tile until all remaining grout film is gone. If buffing does not remove all the film, try using a coarser cloth, like burlap, or even an abrasive pad to remove stubborn grout film.

22 Remove the caulking backer rods from the tile joints, then fill the joints with caulk that is tinted to match the grout color closely. The caulk will allow for some expansion and contraction of the tiled surface, preventing cracking and buckling.

23 Apply grout sealer to the grout lines using a sash brush or small sponge brush. Avoid spilling over onto the tile surface with the grout sealer. Wipe up any spills immediately.

24 OPTION: After one to three weeks, seal the tiled surface with tile sealer, following the manufacturer's application directions. A paint roller with an extension pole is a good tool for applying tile or concrete sealer.

Brand-new concrete steps give a fresh, clean appearance to your house. And if your old steps are unstable, replacing them with concrete steps that have a non-skid surface will create a safer living environment.

Concrete Building Projects
Concrete Steps

Replacing old, cracked or damaged steps with sturdy concrete steps will make your home a safer place for your family and for visitors.

Designing steps requires some math work and a fair amount of trial and error. The basic idea is to come up with a design that fits the space while following practical safety guidelines. You can adjust different elements, such as the landing depth, and the height and depth of each step, as long as the steps meet recommended safety guidelines. Sketching your plan on paper will make the job easier.

Before demolishing your old steps, measure them to see if they meet safety guidelines. If so, you can use them as a reference for your new steps. If not, start from scratch so your new steps do not repeat any design errors.

Step Safety Guidelines:

• *Landing depth* should be at least 12" more than the width of the door; *step treads* should be between 10" and 12" deep; *step risers* should be between 6" and 8" in height.

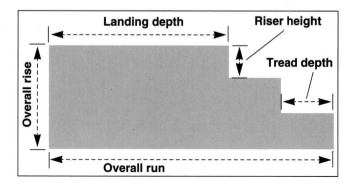

How to Design Steps

1 Attach a mason's string to the house foundation, 1" below the bottom of the door threshold. Drive a stake where you want the base of the bottom step to fall. Attach the other end of the string to the stake and use a line level to level it. Measure the length of the string—this distance is the overall depth, or *run*, of the steps.

2 Measure down from the string to the bottom of the stake to determine the overall height, or *rise*, of the steps. Divide the overall rise by the estimated number of steps. The rise of each step should be between 6" and 8". For example, if the overall rise is 21" and you plan to build three steps, the rise of each step would be 7" (21 divided by 3), which falls within the recommended safety range for riser height.

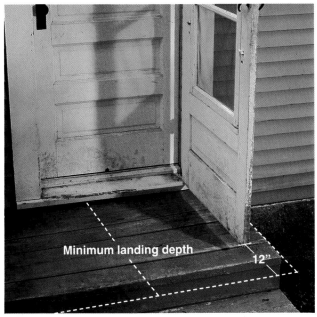

3 Measure the width of your door and add 12"; this number is the minimum depth you should plan for the landing area of the steps. The landing depth plus the depth of each step should fit within the overall run of the steps. If necessary, you can increase the overall run by moving the stake at the planned base of the steps away from the house, or by increasing the depth of the landing.

4 Sketch a detailed plan for the steps, keeping these guidelines in mind: each step should be 10" to 12" deep, with a riser height between 6" and 8", and the landing should be 12" deeper than the swing radius (width) of your door. Adjust the parts of the steps as needed, but stay within the given ranges. Creating a final sketch will take time, but it is worth doing carefully.

1 Remove or demolish existing steps; if the old steps are concrete, set aside the rubble to use as fill material for the new steps. Wear protective gear, including eye protection and gloves, when demolishing concrete.

2 Dig two 12"-wide trenches for the footings, at the required depth (pages 66 to 67). Locate the trenches perpendicular to the foundation, spaced so the footings will extend 3" wider than the outside edges of the finished steps. Install steel rebar grids (page 69) for reinforcement. Since they will support poured concrete, simply strike off the footings level with the ground, rather than building forms.

3 Mix the concrete and pour the footings. Smooth the concrete with a screed board (page 28). You do not need to float the surface afterwards.

4 When bleed water disappears (page 29), insert 12" sections of rebar 6" into the concrete, spaced at 12" intervals and centered side to side. Leave 1 ft. of clear space at each end.

5 Let the footings cure for two days, then excavate the area between them to 4" deep. Pour in a 5"-thick layer of compactible gravel subbase and tamp until it is level with the footings.

Make slopes a minimum of ⅛" per foot

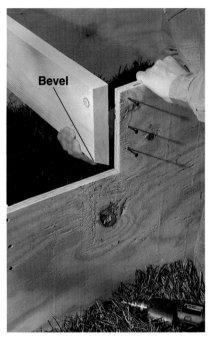

Bevel

6 Transfer the measurements for the side forms from your working sketch onto ¾" exterior-grade plywood. Cut out the forms along the cutting lines, using a jig saw. Save time by clamping two pieces of plywood together and cutting both side forms at the same time. Add a ⅛" per ft. back-to-front slope to the landing part of the form.

7 Cut form boards for the risers to fit between the side forms. Bevel the bottom edges of the boards when cutting to create clearance for the float at the back edges of the steps. Attach the riser forms to the side forms with 2" deck screws.

Cleats

Riser support

8 Cut a 2 × 4 to make a center support for the riser forms. Use 2" deck screws to attach cleats to the riser forms, then attach the support to the cleats. Check to make sure all corners are square.

9 Cut an isolation board (page 21) and glue it to the house foundation at the back of the project area. Set the form onto the footings, flush against the isolation board. Add 2 × 4 bracing arms to the sides of the form, and attach them to cleats on the sides and to stakes driven into the ground.

(continued next page)

10 Fill the form with clean fill (broken concrete or rubble). Stack the fill carefully, keeping it 6" away from the sides, back, and top edges of the form. Shovel smaller fragments onto the pile to fill the void areas.

11 Lay pieces of #3 metal rebar on top of the fill pile for reinforcement. Space the pieces at 12" intervals, and attach them to bolsters with wire to keep them from moving when the concrete is poured. Keep rebar at least 2" below the tops of the form.

12 Mix the concrete and start pouring the steps one at a time, beginning with the bottom step. Settle the concrete and smooth it with a screed board (pages 27 to 28). Press a piece of #3 rebar 1" down into the "nose" of each tread for reinforcement.

13 Float the steps with a wood float (page 29), working the front edge of the float underneath the beveled edge at the bottom of each riser form.

14 Pour concrete in the forms for the remaining steps and the landing, keeping an eye on the poured concrete as you work. Stop to float any concrete as soon as the bleed water disappears (page 29). Press rebar into the nose of each step.

OPTION: For railings with mounting plates that attach to sunken J-bolts, install the bolts before the concrete sets. Otherwise, choose railings with surface-mounted hardware (see step 16) that can be attached after the steps are completed.

15 Once the concrete has set, shape the steps and the landing with a step edger, then float and sweep the surface with a stiff-bristled broom (pages 29 to 31).

Mounting plate

16 Let the concrete cure for one week, then remove the form and install a handrail (check with your local building department for handrail requirements). Backfill the area around the base of the steps, and seal the concrete if desired.

Footings are required by most building codes for concrete or brick and block structures that adjoin other permanent structures. Structures taller than 3 ft. normally require a footing (pages 74 to 75).

Concrete Building Projects
Footing

Footings provide a stable, level base for brick, block, or poured concrete structures. They distribute the weight of the structure evenly, prevent sinking, and keep structures from moving during freezing and thawing.

The required depth of a footing is usually determined by the "frost line," which varies by region. The frost line is the point nearest ground level where the soil does not freeze. In colder climates, it is likely to be 48" or deeper. Frost footings (footings designed to keep structures from moving during freezing temperatures) should be built 12" deeper than the frost line for the area. Check with your local building department to find the frost line depth for your area.

Tips for Planning:

• Describe the structure you intend to build to your local building inspector to find out if it requires a footing, and if the footing needs reinforcement.

• Keep footings separate from adjoining structures by installing an isolation board (page 21).

• For smaller poured concrete projects, consider pouring the footing and the structure as one unit.

• In some cases, slab footings can be used, as long as the subbase provides adequate drainage.

• Footings generally are not built with a slope.

Options for Forming Footings

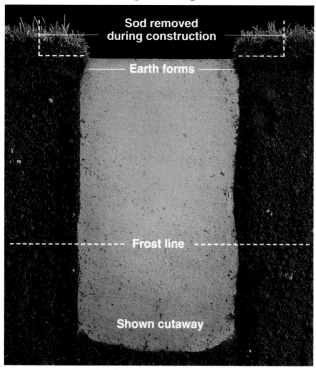

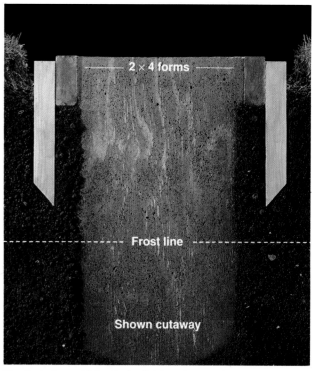

For poured concrete, use the earth as a form. Strip sod from around the project area, then strike off the concrete with screed board resting on the earth at the edges of the top of the trench.

For brick and block, build level 2 × 4 forms. Rest the screed board on the frames when you strike off the concrete to create a flat, even surface for stacking masonry units.

Tips for Building Footings

Make footings twice as wide as the width of the wall or structure they will support. They also should extend at least 12" past the ends of the project area.

Add tie-rods if you will be pouring concrete over the footing. After the concrete sets up, press 12" sections of rebar 6" into the concrete. The tie-rods will anchor the footing to the structure it will support.

How to Pour a Footing

1 Make a rough outline of the footing, using a rope or hose. Outline the project area with stakes and mason's strings (page 22).

2 Strip away sod 6" outside the project area on all sides, then excavate the trench for the footing to a depth 1 ft. below the frostline.

3 Build and install a 2 × 4 form frame for the footing, aligning it with the mason's strings (page 24). Stake the form in place, and adjust to level.

VARIATION: If your project abuts another structure, like a house foundation, slip a piece of fiber board into the trench to create an isolation joint between the footing and the structure (page 21).

4 Make two #3 rebar grids to reinforce the footing. For each grid, cut two pieces of #3 rebar 8" shorter than the length of the footing, and two pieces 4" shorter than the depth of the footing. Bind the pieces together with 16-gauge wire, forming a rectangle. Set the rebar grids upright in the trench, leaving 4" of space between the grids and the walls of the trench. Coat the inside edge of the form with vegetable oil.

5 Mix and pour concrete into the trench, filling it up to the top edges of the forms (pages 26 to 29). Strike off the concrete, using a 2 × 4 as a screed board. Add tie-rods if needed (page 67). Float the surface.

6 Cure the concrete for at least two days (one week is better) before you build on the footing. Remove the forms and backfill the area around the edges of the footing (page 33).

Sealing & Maintaining Concrete

Protect concrete that is exposed to heavy traffic or constant moisture by sealing it with a clear concrete sealer. In addition to sealer, there are other special-purpose products designed for concrete surfaces. Specially formulated concrete paints, for example, help keep minerals in the concrete from leeching through paint and hardening into a white, dusty film (called efflorescence).

Regular cleaning is an important element of concrete maintenance to prevent deterioration from oils and deicing salts. Use concrete cleaner products for scheduled cleanings, and special solutions (page 106) for specific types of stains.

Everything You Need:

Tools: paint brush, paint roller and tray, dust brush and pan, caulk gun, paint pad.

Materials: masonry paint, paint thinner, repair caulk, sealer, concrete recoating product.

Use waterproof concrete paint to paint concrete surfaces. Concrete paint is formulated to resist chalking and efflorescence. It is sold in several stock colors, or you can have custom colors mixed from a tint base.

Tips for Cleaning & Maintaining Concrete

Clean oil stains by dampening sawdust with paint thinner, then applying the sawdust over the stain. The paint thinner will break apart the stain, allowing the oil to be absorbed by the sawdust. Wipe up with a broom when finished, and reapply as necessary.

Fill control joints in sidewalks, driveways and other concrete surfaces with concrete repair caulk. The caulk fills the joint, preventing water from accumulating and causing damage to the concrete.

Exposed-aggregate sealer is specially formulated to keep aggregate from loosening. It should be applied about 3 weeks after the concrete surface is poured. To apply, wash the surface thoroughly and allow it to dry. Pour some sealer into a roller tray. Make a puddle of sealer in the corner and spread it out evenly with a paint roller and extension pole.

Clear concrete sealer helps create a water-resistant seal on the surface. The more popular concrete sealing products today are acrylic based and do not attract dirt. Some types of sealer, like the product shown (right), als help the concrete cure evenly.

Masonry recoating products are applied like paint, but they look like fresh concrete when they dry. They are used frequently to improve the appearance of walls, although they generally have little value as waterproofing agents.

Create a formal setting around your house by building with brick and block. The brick steps and brick veneer on the house above give the home a distinctive, easy-to-maintain appearance.

Building with Brick & Block

A brick or block building project can transform an ordinary house into a striking home. The colors and textures of brick and block structures work with the other materials on the outside of your home and, when applied with good design principles, they give the entire home and yard a sense of balance and composition. At the same time, brick and block building projects, like those shown on the next page, can add practical features that increase the enjoyment of your house for you and your family.

Careful planning and a good design will help you build a project that makes sense for your house, your yard, and your budget. Plan your project to avoid unnecessary cutting of masonry units. Also, consider the size and style of the project.

Brick and decorative block colors and textures vary widely by region, and change frequently to reflect current design trends. Product lines are often discontinued without notice, so always buy extra materials for future use in repairing or updating the structure.

This section shows:

- Brick & Block Basics (pages 78 to 81)
- Mixing & Throwing Mortar (pages 82 to 83)
- Laying Brick & Block (pages 84 to 91)
- Building with Brick & Block: A Step-by-Step Overview (pages 92 to 93)
- Brick-paver Step Landing with Planter Option (pages 94 to 99)
- Brick Veneer (pages 100 to 105)

Common Brick & Block Projects

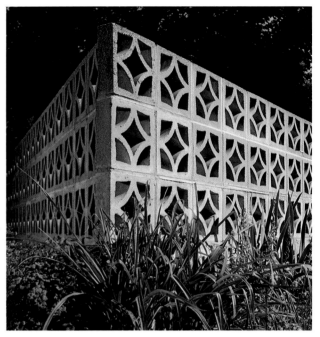

A garden wall is an excellent project for beginners. If the wall is under 3 ft. in height and is not tied to another structure, a frost footing is unnecessary in most cases—check local building codes. Simply pour a slab footing (pages 66 to 69), and build on top of it. A small project like this garden wall lets you take your time as you learn the techniques. Pages 88 to 91.

A decorative-block screen creates a visual barrier without completely blocking light. The open structure of decorative block also allows full air circulation, making screens like the one shown above excellent structures for accenting gardens and enclosing utility areas. Pages 92 to 93.

A brick-paver step landing and planter are effective examples of combining projects to achieve results that are both attractive and useful. The paver landing was set into mortar directly over the existing sidewalk. Pages 94 to 99.

Brick veneer usually is applied to houses during new construction as an attractive design element. But if your foundation walls are in good shape, you can install it as a retrofit project to update the appearance of your home. If the foundation is questionable in any way, have a contractor examine it before you start a veneer project. Pages 100 to 105.

Tips for Planning a Brick or Block Project

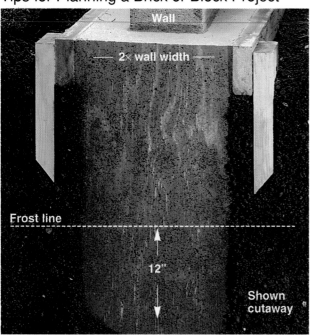

Build a frost footing if the structure is more than 3 ft. tall or if it is tied to another permanent structure (see photo, page opposite, bottom left). Frost footings should be twice as wide as the wall, and should extend about 12" past the frost line. See pages 66 to 69.

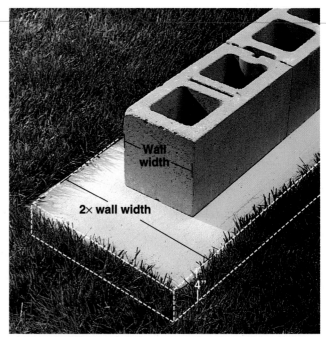

Pour a concrete slab for brick or block building projects that are free-standing and under 3 ft. tall. The slab should be twice the width of the wall, flush with ground level, and about 4" thick. The techniques for pouring a concrete slab are very similar to those for pouring walkways. See pages 34 to 39 and 66 to 69.

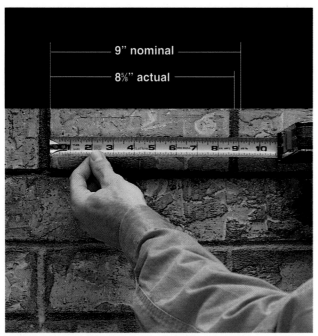

Do not add mortar joint thickness to total project dimensions when planning brick and block projects. The actual sizes of bricks and blocks are ⅜" smaller than the nominal size to allow for ⅜"-wide mortar joints. For example, a 9" (nominal) brick has an actual dimension of 8⅝", so a wall that is built with four 9" bricks and ⅜" mortar joints will have a finished length of 36" (4 × 9").

Test project layouts using ⅜" spacers between masonry units to make sure the planned dimensions work. If possible, create a plan that uses whole bricks or blocks, eliminating extensive cutting.

Select a construction design that makes sense for your project. There are two basic methods used in stacking brick or block. Structures that are only one unit wide are called *single wythe,* and are typically used for projects like brick barbecues or planters, and for brick veneers. *Double-wythe* walls are two units wide and are used as free-standing brick structures. Most concrete-block structures are single wythe.

Build end-support pillars at the free ends of structures, especially when building with nonstructural block, like the decorative block shown above. This wall support pillar uses 6 × 6" concrete blocks.

Keep structures as low as you can. Local codes require frost footings and additional reinforcement for permanent walls or structures that exceed maximum height restrictions—usually 3 ft. or 4 ft. To be on the safe side, try to design walls that are less than 3 ft. tall.

Add a lattice panel or another decorative element to permanent walls to create greater privacy without adding structural reinforcement to the permanent structure.

Common Brick & Block Styles

Common types of brick and block used for residential construction include: decorative block (A) available colored or plain, decorative concrete pavers (B), fire brick (C), standard 8 × 8 × 16" concrete block (D), half block (E), combination corner block (F), queen-sized brick (G), standard brick pavers (H), standard building bricks (I), and limestone wall cap (J).

Common Brick & Block Building Patterns

Running bond is a popular pattern for walls and other vertical projects because the staggered joints create an interlocking effect that increases strength. The pattern also is popular for paving projects (inset).

Stack bond provides a clean geometric look, but is not as strong as a running bond pattern. In most cases, reinforcement should be used when building with a stack bond pattern. When used for paving projects, the pattern is called *jack-on-jack* (inset).

Options for Reinforcing Brick & Block Structures

For double-wythe brick projects, use metal ties between wythes for reinforcement. Insert ties directly into the mortar 2 to 3 ft. apart, every third course. Insert metal rebar into the gap between wythes every 4 to 6 ft. (check local building codes). Insert ¾"-diameter plastic tubing between wythes to frame in the mortar. Pour a thin mixture of mortar between the wythes to improve the strength of the wall.

For block projects, fill the empty spaces (cores) of the block with thin mortar. Insert sections of metal rebar into the mortar to increase vertical strength. Always check with your local building inspector to determine reinforcement requirements, if any.

Provide horizontal reinforcement on brick or block walls by setting metal reinforcing strips into the mortar every third or fourth course. Metal reinforcing strips, along with most other reinforcing products, can be purchased from brick and block suppliers. Overlap the ends of metal strips where they meet.

Brick and block are easy to work with if you have the right tools and use good techniques. For small jobs, cutting can be done with a mason's chisel and hammer (above). For larger cutting projects, use a circular saw with a masonry-cutting blade to score the bricks or blocks (page 80), or have the masonry units cut to size at a brick yard. TIP: To avoid cracking bricks when cutting, set them on a bed of sand.

Brick & Block Basics

Before starting a brick and block project, familiarize yourself with the handling skills for these materials, shown on the next few pages.

Cutting brick or block is a basic skill you will need to learn before beginning any brick or block building project. Make practice cuts on a few samples to determine how the materials respond to different cutting techniques. Bricks and blocks vary in density and the type of materials used to make them, which greatly affects how they respond to cutting. Testing the water absorption rate (next page) is a good way to evaluate the density of the masonry unit.

In addition to cutting, practice your brick and block stacking techniques (next page) to get a feel for working with the type of masonry unit you will be using. Be sure to buy extra brick and block for every project so you can practice cutting and stacking.

Use bricks tongs to carry several bricks at one time, saving time and preventing damage to bricks. Tongs are sold at brick yards and building centers.

Tips for Working with Brick

Make practice runs on a 2 × 4 to help you perfect your mortar-throwing (pages 82 to 83) and bricklaying techniques. You can clean and reuse the bricks to make many practice runs if you find it helpful, but do not reuse the bricks in your actual project—old mortar can impede bonding.

Test the water absorption rate of bricks to determine density. Squeeze out 20 drops of water in the same spot on the surface of the brick. If the surface is completely dry after 60 seconds, predampen the bricks with water before you lay them to prevent the dry brick from leeching the moisture out of the mortar before it has a chance to set properly.

Tips for Marking Bricks & Blocks

Use a T-square and pencil to mark several bricks for cutting. Make sure the ends of the bricks are all aligned.

Mark angled cuts by dry-laying the project (as shown with concrete pavers above) and setting the brick or block in position, allowing for ⅜" mortar joints when necessary. Mark cutting lines with a pencil, using a straightedge where practical to ensure that cutting lines are straight.

How to Score & Cut Brick

Use a mason's chisel and hammer to score all four sides of the brick when cuts fall over the web area, and not over the core (as shown on page 78, top). Tap on the mason's chisel to leave scored cutting marks about ⅛" to ¼" deep.

Use a circular saw with a masonry-cutting blade to gang-cut bricks or blocks along cutting lines, ensuring uniformity and speeding up the process. Clamp the bricks securely at each end with a pipe clamp or bar clamp, making sure the ends are aligned. Split the bricks, using a mason's chisel and a hammer. NOTE: Wear eye protection when cutting masonry units.

How to Angle-cut Brick

1 Mark a cutting line on the brick, then score a straight line in the waste area of the brick about ⅛" away from the starting point of the cutting line, perpendicular to the side of the brick.

Pivot point

Cutting marks

2 To avoid ruining the brick, you will need to make gradual cuts. Keep the chisel stationary at the point of the first cut, pivot it slightly, then score and cut again. It is important to keep the pivot point of the chisel at the edge of the brick. Repeat until all of the waste area is removed.

How to Cut Brick with a Brick Splitter

1 The brick splitter is a tool that makes accurate, consistent cuts in brick and pavers. It is a good idea to rent one if your project requires many cuts. To use the brick splitter, first mark a cutting line on the brick, then set the brick on the table of the splitter, aligning the cutting line with the cutting blade on the tool.

2 Once the brick is in position on the splitter table, pull down sharply on the handle. The cutting blade on the splitter will cleave the brick along the cutting line. TIP: For efficiency, mark cutting lines on several bricks at the same time (see page 79).

How to Cut Concrete Block

1 Mark cutting lines on both faces of the block, then score ⅛" to ¼"-deep cuts along the lines with a circular saw equipped with a masonry blade.

2 Use a mason's chisel and hammer to split one face of the block along the cutting line. Turn the block over and split the other face.

OPTION: Cut half blocks from combination corner blocks. Corner blocks have preformed cores in the center of the web. Score lightly above the core, then rap with a mason's chisel to break off half blocks.

Mixing & Throwing Mortar

Watching a professional bricklayer at work is an impressive sight, even for do-it-yourselfers who have accomplished numerous masonry projects successfully. The mortar practically flies off the trowel and always seems to end up in perfect position to accept the next brick or block.

Although "throwing mortar" is an acquired skill that takes years of practice to perfect, a beginning bricklayer can use the basic techniques successfully with just a little practice.

The first critical element to handling mortar effectively is the mixture. If the mortar is too thick, it will fall off the trowel in a heap, not in the smooth line that is the goal of throwing mortar. If it is too watery, the mortar is impossible to control and will simply make a mess. Experiment with different water ratios until you find the perfect mixture that will cling to the trowel just long enough for you to deliver it in a controlled, even line that holds its shape after settling. Take careful notes on how much water you add to each batch, then write down the best mixture for your records.

Do not mix more mortar than you can use in 30 minutes. Once mortar begins to set up, it is difficult to work with and yields poor results.

Throwing mortar is a quick, smooth technique that requires practice. Load the trowel with mortar (steps 2 and 3, next page), then position the trowel a few inches above the starting point. In one motion, begin turning your wrist over and quickly move the trowel across the surface to spread mortar consistently. Proper mortar-throwing results in a rounded line about 2½" wide and about 2 ft. long.

Everything You Need:

Tools: trowel, hoe, shovel.

Materials: mortar mix, mortar box, plywood.

How to Mix & Throw Mortar

1 Empty mortar mix into a mortar box and form a depression in the center. Add about three-fourths of the recommended amount of water into the depression, then mix it in with a masonry hoe. Do not overwork the mortar. Continue adding small amounts of water and mixing until the mortar reaches the proper consistency. Do not mix too much mortar at one time—mortar is much easier to work with when it is fresh.

2 Set a piece of plywood on blocks at a convenient height, and place a shovelful of mortar onto the surface. Slice off a strip of mortar from the pile, using the edge of your mason's trowel. Slip the trowel point-first under the section of mortar and lift up.

3 Snap the trowel gently downward to dislodge excess mortar clinging to the edges. Position the trowel at the starting point, and "throw" a line of mortar onto the building surface (see technique photos, previous page). A good amount is enough to set three bricks. Do not get ahead of yourself. If you throw too much mortar, it will set before you are ready.

4 "Furrow" the mortar line by dragging the point of the trowel through the center of the mortar line in a slight back-and-forth motion. Furrowing helps distribute the mortar evenly.

Laying Brick & Block

Patience, care, and good brick-laying techniques are the key elements to building brick and block structures with a professional appearance.

This section features two demonstration projects, a concrete block wall and a brick wall, that show the two basic methods you can follow when building with brick and block. In the block project, the courses are built one at a time; in the brick project, the ends are constructed, then the interior bricks are filled in.

Make sure to have a sturdy, level building surface before you start.

Everything You Need:

Tools: gloves, trowel, chalkline, level, mason's string.

Materials: mortar mix, 6 × 6" concrete block.

"Buttering" is a term used to describe the process of applying mortar to a brick or block before adding it to the structure being built. For bricks, apply a heavy layer of mortar to one end, then cut off the excess with a trowel. For concrete blocks (inset), apply a mortar line on each of the flanges at one end.

How to Lay Concrete Block

1 Dry-lay the first course, leaving a ⅜" gap between blocks (page 88). Draw reference lines on the concrete base to mark the ends of the row, extending the lines well past the edges of the block. Use a chalkline to snap reference lines on each side of the base, 3" from the blocks. These reference lines will serve as a guide when setting the blocks into mortar.

2 Dampen the base slightly, then mix mortar and throw and furrow two mortar lines (page 83) at one end to create a mortar bed for the combination corner block. Dampen porous blocks (page 79) before setting them into the mortar beds.

3 Set a combination corner block (page 76) into the mortar bed. Press it into the mortar to create a ⅜"-thick bed joint. Hold the block in place and cut away the excess mortar (save excess mortar for the next section of the mortar bed). Check the block with a level to make sure it is level and plumb. Make any necessary adjustments by rapping on the high side with the handle of a trowel. Be careful not to displace too much mortar.

4 Drive a stake at each end of the project and attach one end of a mason's string to each stake. Thread a line level onto the string and adjust the string until it is level and flush with the top of the corner block. Throw a mortar bed and set a corner block at the other end. Adjust the block so it is plumb and level, making sure it is aligned with the mason's string.

5 Throw a mortar bed for the second block at one end of the project: butter one end of a standard block (page 84) and set it next to the corner block, pressing the two blocks together so the joint between them is ⅜" thick. Tap the block with the handle of a trowel to set it, and adjust the block until it is even with the mason's string. Be careful to maintain the ⅜" joint.

6 Install all but the last block in the first course, working from the ends toward the middle. Align the blocks with the mason's string. Clean excess mortar from the base before it hardens.

(continued next page)

7 Butter the flanges on both ends of a standard block for use as the "closure block" in the course. Slide the closure block into the gap between blocks, keeping the mortar joints an even thickness on each side. Align the block with the mason's string.

8 Apply a 1"-thick mortar bed for the half block at one end of the wall, then begin the second course with a half block (page 76).

9 Set the half block into the mortar bed with the smooth surfaces facing out. Use the level to make sure the half block is plumb with the first corner block, then check to make sure it is level. Adjust as needed. Install a half block at the other end.

VARIATION: If your wall has a corner, begin the second course with a full-sized end block that spans the vertical joint formed where the two walls meet. This layout creates and maintains a running bond for the wall (page 76).

10 Attach a mason's string for reference, securing it either with line blocks (page 76) or a nail. If you do not have line blocks, insert a nail into the wet mortar at each end of the wall, then wind the mason's string around and up to the top corner of the second course as shown above. Connect both ends and draw the mason's string taut. Throw a mortar bed for the next block, then fill out the second course, using the mason's string as a reference line.

11 Every half-hour, tool the fresh mortar joints with a jointing tool and remove any excess mortar. Tool the horizontal joints first, then the vertical joints. Cut off excess mortar, using a trowel blade. When the mortar has set, but is not too hard, brush any excess mortar from the brick faces. Continue building the wall until it is complete.

OPTION: When building stack bond walls with vertical joints that are in alignment, use wire reinforcing strips in the mortar beds every second or third course (or as required by local codes) to increase the strength of the wall. The wire should be completely embedded in the mortar. See page 77 for other block wall reinforcing options.

12 Install a wall cap (page 76) on top of the wall to cover the empty spaces and create a finished appearance. Set the cap pieces into mortar beds, then butter an end with mortar. Level the cap, then tool to match the joints in the rest of the wall.

Tips for Planning a Brick Wall

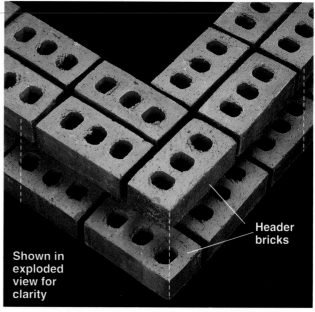

Dry-lay the first course of the wall to test your layout. If you are building a double-wythe wall, the wythes should be ¾" to 1" apart for most walls. Use a chalk-line to outline the location of the wall on the slab. Draw pencil lines on the slab to mark the ends of the bricks. Test-fit spacing with a ⅜"-diameter dowel, then mark the location of joint gaps to maintain a reference after the spacers are removed.

To build corners for a double-wythe wall, lay a header brick at the end of two parallel wythes, then reverse the position of the header brick on the next course so it is perpendicular to the first header brick. Continue alternating the pattern all the way to the top of the wall.

How to Build a Double-wythe Brick Wall

1 Dampen the concrete slab or footing with water, and dampen the bricks or blocks if they are porous (page 79, top). Mix mortar and throw a mortar bed (page 83) for the first two bricks of one of the wythes. Butter the inside end of the first brick, then press the brick into the mortar, creating a ⅜" mortar bed. Cut away excess mortar.

2 Plumb the face of the end brick, using a level. Tap lightly with the handle of the trowel to correct the brick if it is not plumb. Level the brick end to end. Butter the end of a second brick (page 84), then set it into the mortar bed, pushing it toward the first brick to create a joint of ⅜".

3 Butter and place a third brick, using the chalklines as a general reference, then using a level to check for level and plumb. Adjust any bricks that are not aligned by tapping lightly with the trowel handle.

4 Lay the first three bricks for the other wythe, parallel to the first wythe. Use a level to level the wythes, and make sure the end bricks and mortar joints align. Fill the gaps between the wythes at each end with mortar.

5 Cut a half brick (page 76), then throw and furrow a mortar bed (pages 82 to 83) for a half brick on top of the first course. Butter the end of the half brick with mortar, then set the half brick in the mortar bed, creating a ⅜" joint. Cut away excess mortar. Check with a level to make sure bricks are plumb and level.

6 Add more bricks and half bricks to both wythes of the corner until you finish the fourth course. Check the corner frequently with a level to make sure it is level and plumb. Align bricks with the reference lines (page 88).

(continued next page)

How to Build a Double-wythe Brick Wall (continued)

7 Check the spacing of the end bricks with a straight-edge. Properly spaced bricks will form a straight line when you place the straightedge over the stepped end bricks. If bricks are not in alignment, do not move those bricks already set. Try to compensate for the problem gradually as you fill in the middle (field) bricks by slightly reducing or increasing the spacing between the joints.

8 Every 30 minutes, stop laying bricks and smooth out all the untooled mortar joints with a jointing tool. Do the horizontal joints first, then the vertical joints. Cut away any excess mortar pressed from the joints, using a trowel. When the mortar has set, but is not too hard, brush any excess mortar from the brick faces.

Line block

9 Build the opposite end of the wall with the same methods as the first, using the chalklines as a reference. Stretch a mason's string between the two ends to establish a flush, level line between ends— use a line block to secure the string (page 11). Tighten the string until it is taut. Begin to fill in the field bricks (the bricks between corners) on the first course, using the mason's string as a guide.

10 Lay remaining bricks in the field course. The last brick, called the closure brick, should be buttered at both ends. Center the closure brick between the two adjoining bricks, then set in place with the trowel handle. Fill in the first three courses of each wythe, moving the mason's string up one course after completing each course.

Metal wall tie

11 In the fourth course, set metal wall ties into the mortar bed of one wythe and on top of the brick adjacent to it. Space the ties 2 ft. apart, every three or four courses. For added strength, set metal rebar into the cavities between the wythes and fill with thin mortar (page 77).

12 Fill in the remaining courses. Cut away excess mortar. Check the mason's string frequently for alignment, and use a level to make sure the wall stays plumb and level.

13 Lay a furrowed mortar bed on the top course, and place a wall cap on top of the wall to cover empty spaces and provide a finished appearance. Remove any excess mortar. Make sure the cap blocks are aligned and level. Fill the joints between cap blocks with mortar.

Building with Brick & Block: A Step-by-Step Overview

Building with brick and block requires a higher level of precision than building with poured concrete. Poured concrete is fairly forgiving, but bricks and blocks must be aligned precisely and joined with perfectly uniform mortar joints.

The decorative-block screen shown here is a continuation of the poured concrete project shown on pages 32 to 33, and it demonstrates most of the techniques needed for working with brick and block. Use decorative-block screens to hide garbage cans or air conditioning units, or as simple garden walls. See pages 74 to 77 for information on brick and block styles and construction options.

1 Make a dry run of the first course of blocks to make sure the layout meets your needs. Use spacers to represent mortar joints. Good planning will minimize the need to cut bricks or blocks, and will speed up your work. Once you have established the layout to your satisfaction, mark layout lines on the concrete base (page 84).

2 With layout lines clearly marked and your brick or block supply nearby, mix the mortar and throw a mortar bed at one corner or end (pages 82 to 83). Set the first block or brick into the bed. Tap the block with the handle of your trowel to set it into the mortar, and check with a level to make sure the block is level and plumb (pages 84 to 85).

3 Complete the first course, maintaining a uniform mortar bed and joint thickness. For the project shown, the first block for the end pillar was also laid. Begin the second course, following your stacking pattern (pages 85 to 87). OPTION: Establish the layout by building corners first, then add blocks or bricks between corners (pages 88 to 91).

4 Smooth out fresh mortar joints (called tooling) within 30 minutes. Use a mortar jointing tool, like the V-shaped mortar tool shown above, to create clean joints with a consistent look. Tool the horizontal joints first, then tool the vertical joints (page 87). Clean off any excess mortar with a brush after it hardens slightly.

5 Continue laying courses of brick or block, using a mason's string and a level to check the alignment (pages 86 to 87).

6 Add reinforcement to the structure as needed (page 77). There are several options for reinforcing brick and block structures; the option you select depends on the type of building materials you are using, and on building code regulations. For this project, we embedded metal wire reinforcing strips in the mortar beds for the entire third course.

7 Add a wall cap to increase the strength of the structure, to cover voids in some types of brick or block, and to provide a decorative finish to the project (pages 87 and 91). Let the mortar cure uncovered for at least one week before subjecting it to stress.

Brick pavers and standard building bricks are combined to create an integrated landing area and planter in this project. Because neither the planter nor the landing are constructed with frost footings, they must not be attached, so they can move independently without cracking.

Brick & Block Building Projects

Brick-paver Step Landing with Planter Option

The entry area is the first detail that visitors to your home will notice. Create a memorable impression by building a brick-paver step landing that gives any house a more formal appearance. Add a special touch to the step landing by building a permanent planter next to it, using matching brick.

In many cases, a paver landing like the one shown here can be built directly over an existing sidewalk. Make sure the sidewalk is structurally sound and free from major cracks. If adding an adjoining structure, like a planter, create a separate building base and be sure to include isolation joints so the structure is not connected to the landing area or to the house.

Tips for Planning:

• Use pavers made from solid brick or concrete for all paving projects. Pavers are denser than standard building bricks to allow them to withstand high traffic volumes.

• Adjust the width of the mortar joints on mortar-set pavers to help the project fit into a defined space. Do not make joints less than ¼" or more than ¾".

• Pay close attention to scale when combining projects. Make sketches to scale on graph paper to help you create balanced projects.

• Repeat design elements, like brick styles or patterns, in both structures to create an integrated appearance.

• Precut pavers or bricks if they will be needed in your project (pages 80 to 81). If you need to cut more than a handful of pavers, consider taking them a brickyard to have them cut by machine.

Variations for Building with Brick Pavers

Consider pattern variations. In addition to the basic running bond and jack-on-jack patterns (page 76), there are many other ways to lay pavers for a more decorative effect. The basketweave and herringbone patterns (above) are two common styles. When considering pattern variations, take into account the amount of cutting that is required by some patterns. Generally, diagonal patterns like the herringbone will require cut pavers all the way around the border.

Stand pavers on end (called "soldier" style) to create a deep border for paving projects. Using soldier-style border pavers is especially effective when you are laying pavers over an existing concrete slab, because the border will hide the exposed edges of the concrete.

Set pavers into sand for backyard landscaping projects, like patios and garden walkways. Using sand-set pavers is a quick, easy method for creating masonry projects, but be prepared to re-set pavers on a semi-regular basis, especially in colder climates. See *Decks & Landscaping* for more information on sand-set pavers.

Pour a concrete slab with the surface about 2" below grade to create a solid, permanent foundation for pavers that are bonded together with mortar. Mortared pavers and a sturdy foundation are desirable for paver projects that will experience high traffic; if the soil and subsoil below the project are unstable; and in areas where extensive freezing and thawing occur, potentially causing heaving of sand-set pavers.

How to Build a Brick-paver Step Landing

1 Dry-lay the pavers onto the concrete surface and experiment with the arrangement to create a layout that uses whole bricks, if possible. Mark outlines for the layout onto the concrete. Attach an isolation board (page 21) to prevent the mortar from bonding with the foundation. Mix a batch of mortar (pages 82 to 83), and dampen the concrete slightly.

2 Throw a bed of mortar (page 83) for three or four border pavers, starting at one end or corner. Level off the bed to about ½" in depth with the trowel.

3 Begin laying the border pavers, buttering an end of each paver with mortar as you would for a brick (page 84). Set pavers into the mortar bed, pressing them down so the bed is ⅜" thick. Cut off excess mortar from the tops and sides of the pavers. Use a level to make sure the pavers are even across the tops, and check mortar joints to confirm that they are uniform thickness.

4 Finish the border section next to the foundation, checking with a level to make sure the row is even in height. Trim off any excess mortar, then fill in the third border section, leaving the front edge of the project open to provide easier access for laying the interior "field" pavers.

5 Apply a ½"-thick bed of mortar between the border pavers in the work area closest to the foundation. Because mortar is easier to work with when fresh, apply the mortar in small sections (no more than 4 sq. ft.).

6 Begin setting pavers in the field area. Check the alignment with a straightedge. Adjust paver height as needed, making sure joints are uniform in width. NOTE: Pavers often are cast with spacing flanges on the sides, but these are for sand-set projects. Use a spacing guide, like a dowel, when setting pavers in mortar. It is not necessary to butter field pavers.

7 Fill in the rest of the pavers to complete the pattern in the field area, applying mortar beds in small sections. Add the final border section. Every 30 minutes add mortar to joints between pavers until it is even with the tops. TIP: To minimize mess when adding mortar, use a mortar bag to deliver the mortar into the joints.

8 Smooth and shape the mortar joints with a jointing tool. Tool the full-width "running" joints first, then tool the joints at the ends of the pavers. Let excess mortar dry until crumbly, then brush it off the surface. Let the mortar set for two or three days, then scrub off any residue with water and a rough-textured rag. Let the mortar cure for one week before allowing foot traffic on the pavers.

How to Build a Brick Planter

1 Lay out the project area with stakes, mason's strings, and a line level (pages 21 to 22). For most brick or block structures, you will need to pour a concrete slab for a foundation. With larger projects, a frost footing often is required; check your local building codes.

2 Excavate the building site, install forms and isolation boards, and pour a concrete base for the project (pages 23 to 25). Let the footing cure for at least three days before building on it. Remove forms, then trim isolation boards so they are level with the tops of adjoining structures, like the landing shown above. TIP: Cover adjoining surfaces for protection.

3 Test-fit the first course of the project, then outline the project on the concrete surface. Dampen the surface slightly, then mix mortar and throw a mortar bed in one corner (pages 82 to 83). Begin laying bricks for the project, buttering the exposed end of each brick before setting it (pages 84 to 85).

Corner return bricks

4 Lay one section of the first course, checking the bricks frequently with a level to make sure the tops are level and even. Lay two corner "return" bricks perpendicular to the end bricks in the first section, and use a level to make sure they are even across the tops.

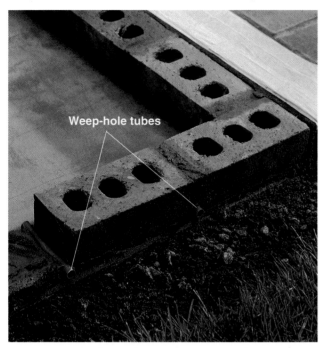

Weep-hole tubes

5 Install weep holes for drainage in the first course of bricks, on the sides farthest away from permanent structures. Cut ⅜"-diameter copper or PVC tubing about ¼" longer than the width of one brick, and set the pieces into the mortar joints between bricks, pressing them into the mortar bed so they touch the footing.

6 Finish building all sides of the first course. Check frequently to make sure the tops of the bricks are level. Lay the second course of bricks, reversing the directions of the corner bricks to create staggered vertical joints if using a running-bond pattern. Fill in brick courses to full height, building up one course at a time (pages 84 to 87).

7 Install cap bricks to keep water from entering the cores of the brick and to enhance the visual effect. (For our project, we used the same brick pavers that we used in the border of the step landing.) Set the cap bricks into a ⅜"-thick mortar bed, buttering one end of each cap brick. Let the mortar cure for one week. Be-

fore filling the planter with dirt, pour a 4" to 6"-thick layer of gravel into the bottom of the planter for drainage, then line the bottom and sides of the planter with landscape fabric to prevent dirt from running into the drainage tubes and causing clogs.

Update the exterior of your house with a layer of brick veneer. Whether it is installed to cover an unattractive foundation or to accent an otherwise ordinary exterior, brick veneer adds color and texture to your house. And when it is installed correctly, it forms an extra layer of protection against the elements.

Brick & Block Building Projects
Brick Veneer

Brick veneer is essentially a brick wall built around the foundation of a house. It is supported by the house with metal wall ties. It is best to use queen-sized bricks for veneer projects because they are thinner than standard building bricks. Even when using queen-sized bricks, veneer structures are quite heavy and subject to a number of local building codes. Always check with an inspector before you start your project.

Special Materials for Project as Shown:

⅜ × 4" lag screws & washers, lead-sleeve style masonry anchors, angle iron for metal shelf supports, PVC roll flashing, corrugated metal wall ties, brickmold for sill extensions, sill-nosing trim.

Tips for Planning:

• Use brick veneer as a decorative accent on the front face or entry area of your house. Installing veneer around an entire house is a time-consuming project complicated by the need to make corners; consult a professional for projects of this scale.

• Check local building codes to learn requirements pertaining to allowable height, reinforcement, use of wall ties, space between veneer and wall sheathing, drainage, and specifications for metal support shelves. A building permit may be required.

• Precut all bricks before you start (steps 2 and 8, following pages).

• Examine the area around the base of your house foundation. It has been a common practice for new-home builders to install a concrete ledge just below ground level as a support base for brick veneer. If your house has no base, attach a metal support shelf to the house foundation.

Options for Installing Brick Veneer

Refresh an old foundation by applying brick veneer from ground level up to the sill plate of the house. The foundation must be free from structural damage.

Full-wall veneer is applied from the foundation to the roof soffit. Because of the extreme weight of full-wall veneer, extensive reinforcement is required. Installation is not generally recommended for do-it-your-selfers; consult a professional.

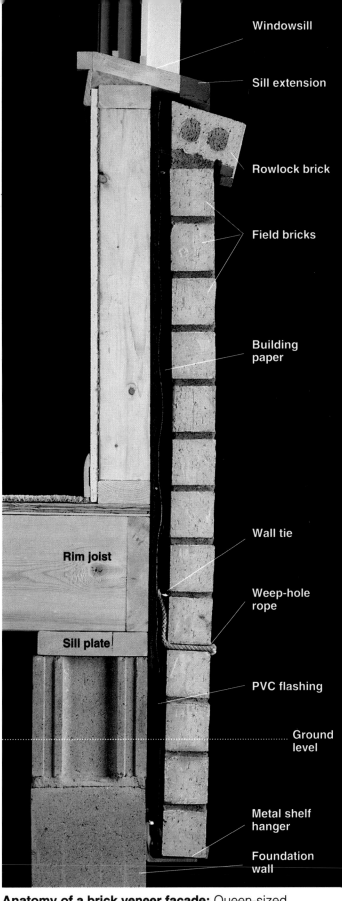

Windowsill

Sill extension

Rowlock brick

Field bricks

Building paper

Wall tie

Weep-hole rope

PVC flashing

Ground level

Metal shelf hanger

Foundation wall

Rim joist

Sill plate

Anatomy of a brick veneer facade: Queen-sized bricks are stacked onto a metal or concrete shelf and connected to the foundation and walls with metal ties. In the project above, "rowlock" bricks are cut to follow the slope of the windowsills, then laid on edge over the top course of bricks.

101

How to Install Brick Veneer

1 The brick veneer in our project is installed over the foundation walls and side walls, up to the bottoms of the windowsills on the first floor of the house. Remove all siding materials in this area before you start. To complete the project, you will need to install a wooden extension onto the windowsill (page 101, right). Before you start laying out the project, cut the wood for the sill extension from treated lumber the same thickness as the sill. Tack the extension onto the sill temporarily.

2 Position a rowlock brick with the top edge against the underside of the sill extension (precut rowlock bricks to follow the slope of the sill and overhang the field brick by about 2"). Use a level to transfer the lowest point on the outside end of the brick onto the wall sheathing. This will establish the height for the top course of veneer bricks in the field area. Extend the line across the entire project area, using a 4-ft. level. Remove the sill extensions.

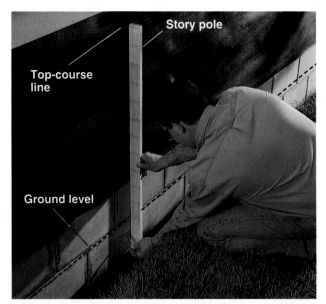

3 Make a story pole by laying bricks next to a piece of 1 × 2" lumber, with a ⅜" gap between bricks for mortar joints, and marking the locations of the bricks onto the pole. Dig a 12"-wide, 12"-deep trench around the foundation. Position the story pole so the top-course line on the sheathing aligns with a brick mark on the pole. Mark a line for the first course onto the foundation wall, below ground level.

4 Extend the mark for the first-course height across the foundation wall, using a level as a guide. Measure the thickness of the metal shelf (usually ¼"), and drill pilot holes into the foundation at 16" intervals along the first-course line, far enough below the line to allow for the thickness of the shelf. Slip 10d nails into the pilot holes to create temporary support for the shelf.

5 Use lag screws and masonry anchors to attach the metal shelf to the foundation wall. Set metal shelf onto the temporary supports; mark the location of the center web of each block onto the vertical face of the shelf; remove the shelf and drill ⅜"-diameter holes at the web marks; set the shelf back onto the temporary supports and outline the predrilled holes onto the blocks; remove the shelf and drill holes for the masonry anchors into the foundation, using a masonry bit; drive the masonry anchors into the holes.

6 Position the metal shelf back on the temporary supports so the predrilled holes in the shelf align with the masonry anchors. Attach the shelf to the foundation wall with ⅜ × 4" lag screws threaded through washers and driven at masonry anchor locations. Allow a ¹⁄₁₆" expansion joint between sections of metal shelf. Remove the temporary support nails.

7 After all sections of the metal shelf are attached, staple 30 mil PVC flashing above the foundation wall so it laps over the metal shelf.

8 Test-fit the first course of veneer bricks on the metal shelf. Work in from the ends toward the middle, using spacers to mark an even ⅜" mortar gap between bricks. Mark the final "closure" brick in the first course for cutting to fit the gap left where the two ends of the run meet. NOTE: Cut bricks will be less visible if you install them one brick in from the end brick of each course.

(continued next page)

9 Begin building up one corner of the veneer, maintaining a gap of ½" to 1" from the wall sheathing. Bricks should not overhang the shelf by more than ⅜". Use a level to make sure the bricks are plumb.

10 Build up both corners to the second course above ground level. Attach line blocks and a mason's string to the top brick at the end of each corner (pages 90 to 91), then fill in the field bricks between the corners so they align with the strings. Smooth fresh mortar joints with a jointing tool every 30 minutes.

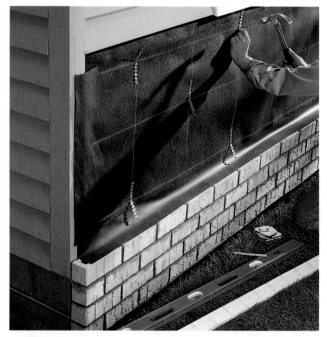

11 Attach another course of PVC flashing to the wall so it covers the top course of bricks, then staple building paper to the wall so it overlaps the top edge of the PVC flashing by at least 12". Mark wall-stud locations on the building paper, using an electronic stud finder if needed.

12 Use the story pole to mark layout lines for the tops of every fifth course of bricks. Attach corrugated metal wall ties to the sheathing where the brick lines meet the marked wall-stud locations.

13 Fill in the next course of bricks, applying mortar directly onto the PVC flashing. At every third mortar joint in this course, tack a 10" piece of ⅜"-diameter cotton rope to the sheathing so it extends all the way through the bottom of the joint, creating a weep hole for drainage. Imbed the metal wall ties in the mortar beds applied to this course.

14 Add courses of bricks, building up corners first, then filling in the field. Imbed the wall ties into the mortar beds as you reach them. Use corner blocks and a mason's string to verify the alignment, and check frequently with a 4-ft. level to make sure the veneer is plumb.

15 Apply a ½"-thick mortar bed to the top course, and begin laying the "rowlock" bricks with the cut ends facing against the wall. Apply a layer of mortar to the bottom of each rowlock brick, then press the brick up against the sheathing, with the top edge following the slopeline of the windowsills.

Sill-nosing trim

16 Finish-nail the sill extensions (step 1) to the windowsills, and nail sill-nosing trim to the siding to cover any gaps above the rowlock course. Fill cores of exposed rowlock blocks with mortar, and caulk any gaps around the veneer with silicone caulk.

Cleaning & Painting Brick & Block

Use a pressure washer to clean large brick and block structures. Pressure washers can be rented from most rental centers. Be sure to obtain detailed operating and safety instructions from the rental agent.

Solvent Solutions for Common Brick & Block Stains

- **Egg splatter:** dissolve oxalic acid crystals in water, following manufacturer's instructions, in a nonmetallic container. Brush onto the surface.

- **Efflorescence:** scrub surface with a stiff-bristled brush. Use a household cleaning solution for surfaces with heavy accumulation.

- **Iron stains:** spray or brush a solution of oxalic acid crystals dissolved in water, following manufacturer's instructions. Apply directly to the stain.

- **Ivy:** cut vines away from the surface (do not pull them off). Let remaining stems dry up, then scrub them off with a stiff-bristled brush and household cleaning solution.

- **Oil:** apply a paste made of mineral spirits and an inert material like sawdust (see page 70).

- **Paint stains:** remove new paint with a solution of trisodium phosphate (TSP) and water, following manufacturer's mixing instructions. Old paint can usually be removed with heavy scrubbing or sandblasting.

- **Plant growth:** use weed killer according to manufacturer's directions.

- **Smoke stains:** scrub surface with household cleanser containing bleach, or use a mixture of ammonia and water.

Check brick and block surfaces annually for stains or discoloration. Most problems are easy to correct if they are treated in a timely fashion. Refer to the information below for cleaning tips that address specific staining problems.

Painted brick and block structures can be spruced up by applying a fresh coat of paint. As with any other painting job, thorough surface preparation and the use of a quality primer are critical to a successful outcome.

Regular maintenance will keep brick and block structures around your house looking their best, helping them last as long as possible.

Tips for Cleaning Masonry
- Always test cleaning solutions on a small area of the surface to evaluate the results.

- Some chemicals and their fumes may be harmful. Be sure to follow manufacturer's safety and use recommendations. Wear protective clothing.

- Soak the surface to be cleaned with water before you apply any solutions. This keeps solutions from soaking in too quickly. Rinse the surface thoroughly after cleaning to wash off any remaining cleaning solution.

Tips for Cleaning Brick & Block Surfaces

Mix a paste made from cleaning solvents (see chart, previous page) and talcum or flour. Apply paste directly to stain, let it dry, then scrape it off with a vinyl or plastic scraper.

Use a nylon scraper or a thin block of wood to clean up spilled mortar that has hardened. Avoid metal scrapers since they are likely to damage masonry surfaces.

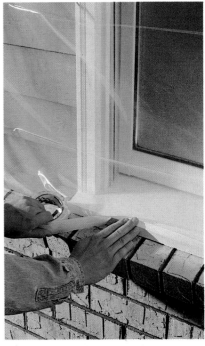

Mask off windows, siding, decorative millwork, and other exposed nonmasonry surfaces before cleaning brick and block. Careful masking is essential if you are using harsh cleaning chemicals like muriatic acid.

Tips for Painting Masonry

Clean mortar joints, using a drill with a wire wheel attachment before applying paint. Scrub off loose paint, dirt, mildew, and mineral deposits so the paint will bond better.

Apply masonry primer before repainting brick or block walls. Primer helps to eliminate stains and prevent problems like efflorescence.

Consult a design professional to help you create a detailed construction plan for your project. Unless you have a great deal of experience with frame carpentry and house construction, you should have a professionally drawn plan before starting any major project, like adding a porch.

Planning a Porch

The adage "form follows function" applies as much to a porch as it does to any other structural design. Begin your planning by listing ways you would like to use your new porch. Are you looking mainly for a comfortable spot where you can contemplate your garden in summer bloom, or do you want a light-filled place where you can enjoy your coffee and the morning paper throughout most of the year?

Next, take stock of your property. Is there room to build around your existing landscape elements without running afoul of local building and zoning restrictions? Is the proposed area exposed to direct sun or wind? Sketching a site drawing will help you assess these factors. Also consider the style and construction of your house. If you are thinking about adding a porch, would it look best with a peaked gable roof that matches your house roof, or would a flat shed roof blend in better? Once you have made some initial decisions about the project that is best for you, enlist the aid of an architect or a designer. He or she can help you develop ideas into detailed design drawings needed to obtain a building permit. Also consult your local building department. The inspectors there can supply you with information about building codes and other requirements that will affect your project.

This section shows:
• Tips for Designing Your Project (pages 109 to 111)
• Working with Inspectors & Building Codes (pages 112 to 113)
• Working Safely (pages 114 to 115)
• Tools & Materials (pages 116 to 117)

Tips for Designing Your Project

Evaluate the planned project site: Note locations of windows, electrical service lines, and any other obstructions that might affect the positioning or design of a porch or a patio structure. For example, if you are interested in building a front porch, note the distance from the front door to any nearby windows—any design you come up with must be large enough to include the window, or small enough to stop short of it. Also assess the building materials used in and around the planned project area. For example, the yellow stucco walls on the house at right would be a natural fit with any clay-based or earth-colored building materials, like terra-cotta patio tile.

Consider the roofline of your house for projects that are attached to the house and include their own roof. The house to the left had just enough second-floor exterior wall space that a porch could be attached to it without the need for tying the porch roof directly to the roof of the house—a project for professionals only. The patio enclosure at the right takes advantage of the low, flat roof on the house expansion by extending the roofline to create the patio-enclosure roof.

(continued next page)

Measure your proposed project area, then draw a scale plan on which you can sketch ideas. Your plan should include relevant features such as shade patterns, trees, and other landscaping details. Also measure the height of door thresholds and the length and height of any walls or buildings adjacent to the proposed project area.

Measure the slope of the proposed building site to determine if you would need to do any grading work. Drive stakes at each end of the area, then tie a mason's string between the stakes. Use a line level to set the string to level. At each stake, measure from the string to the ground: the difference in the distances, when calculated over the distance between stakes, will give you the slope. If the slope is greater than 1" per foot, you may need to regrade the building site: consult a landscape architect.

Measure the roof slope of your house, and try to use the same slope if the project you are planning includes a roof. Hold a carpenter's square against the roofline with the long arm perfectly horizontal. Position the square so the long arm intersects the roof at the 12" mark. On the short arm, measure down to the point of intersection: the number of inches will give you the roof slope in a 12" span. For example, if the top of the square is 4" from the roofline, then your roof slope is 4-in-12.

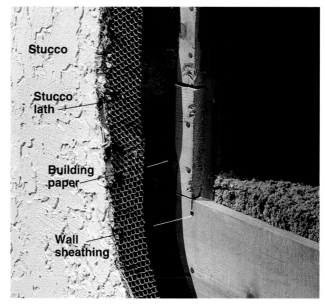

Stucco

Stucco lath

Building paper

Wall sheathing

Identify your siding type:

Stucco (above): If you have stucco siding, plan on a fair amount of work if your project requires you to remove siding—as when installing a ledger board. To remove stucco siding, first score the outline of the area to be removed with a chisel. Next, cut the stucco using a circular saw with a masonry blade. Make multiple passes, increasing the blade depth by ⅛" increments until the blade reaches the stucco lath. Chisel the stucco away inside the cut-out area, using a cold chisel, and cut out the lath with aviator snips.

Lap siding (below): Whether it is wood, aluminum, or vinyl, lap siding is much easier to remove than stucco, and it should not be considered an impediment when planning a project. To remove it, simply set the blade of a circular saw to a cutting depth equal to the siding thickness (usually about ¾") and make straight cuts at the edges of the removal area. Finish the cuts at the corners with a chisel, and remove the siding.

Framing members

Rim joist

Rim joist

Learn about house construction. The model above shows the basic construction of a platform-framed house—by far the most common type of framing today. Pay special attention to locations of rim joists and framing members, since you likely will need to anchor any large porch or patio project to one or both of these elements.

Working with Inspectors & Building Codes

Mark property lines and measure distances from the planned project area to municipal sidewalks or streets before developing a detailed project plan. To avoid future disputes, mark property lines as though they are 1 ft. closer in than they actually are.

Concrete frost footings are required for many porch and patio projects—the frost line is the first point below ground level where freezing will not occur. Frost lines may be 48" or deeper in colder climates. Always build frost footings 1 ft. past the frost line.

Any addition made to your house must comply with local zoning ordinances and building codes that specify where and how you can build.

In most areas, building codes are created by the building or planning department in your municipal government. These departments are staffed by inspectors, who are trained to answer questions, provide information, grant building permits, and make on-site inspections of some building projects. Get to know your local inspectors at the beginning of your project—view them as a helpful resource for a successful project.

The specific types of projects that require building permits varies between localities, but it is safe to say that any major project, like a new porch or a patio enclosure, will require a permit. To be granted a permit, you must present a detailed plan that includes both an elevation drawing and a floor plan (see photos, right). There are some very specific conventions you must follow in creating these drawings, so get assistance if you have not done this kind of work before. You also will be required to pay a permit fee, which likely is based on the projected cost of the project.

In some cases, projects that are designed to occupy space near property borders or municipal sidewalks or streets may require that you get a variance from your local zoning commission. Discuss this possibility with a local inspector.

Project Features Affected by Codes:

• **Construction materials:** Building codes often prescribe minimum sizes for structural members of your project, like deck joists, beams, posts, and ledgers.

• **Fasteners:** Screw sizes and spacing are usually indicated for most parts of the project, from structural elements to roof decking.

• **Total size:** Height, width, and estimated weight of the project must be suitable for the support methods used, and comply with any neighborhood covenants.

• **Footings:** Codes define what, if any, footing requirements affect your project (photo, left).

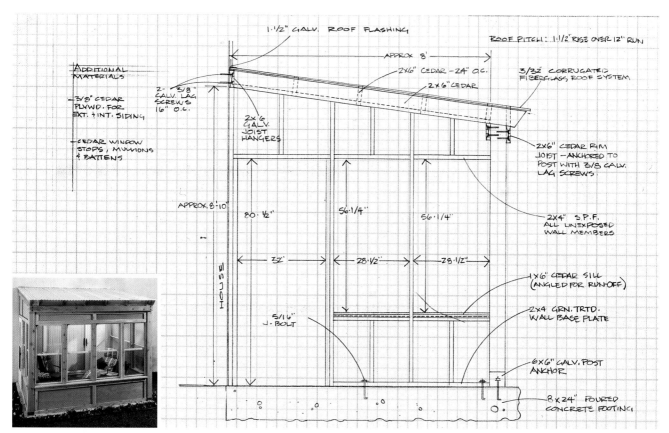

An elevation drawing

- 1.1/2" GALV. ROOF FLASHING
- ROOF PITCH : 1.1/2" RISE OVER 12" RUN
- APPROX 8'
- 2x6" CEDAR - 24" O.C.
- 2x6" CEDAR
- 3/32" CORRUGATED FIBERGLASS ROOF SYSTEM

ADDITIONAL MATERIALS
- 3/8" CEDAR PLYWD. FOR EXT. & INT. SIDING
- CEDAR WINDOW STOPS, MULLIONS & BATTENS

- 2- 3/8" GALV. LAG SCREWS 16" O.C.
- 2x6 GALV. JOIST HANGERS
- APPROX. 8'-10"
- 80.1/2"
- 56.1/4"
- 56.1/4"
- 2x6" CEDAR RIM JOIST - ANCHORED TO POST WITH 3/8 GALV. LAG SCREWS
- 2x4" S.P.F. ALL UNEXPOSED WALL MEMBERS
- HOUSE
- 32"
- 28.1/2"
- 28.1/2"
- 1x6" CEDAR SILL (ANGLED FOR RUN-OFF)
- 5/16" J. BOLT
- 2x4 GRN. TRTD. WALL BASE PLATE
- 6x6" GALV. POST ANCHOR
- 8 x 24" POURED CONCRETE FOOTING

An elevation drawing shows sizes and locations of structural elements such as footing, ledger, studs, beams, rafters, door and window headers, as well as roof pitch and the type of roof covering planned. It also notes specific building materials and fasteners.

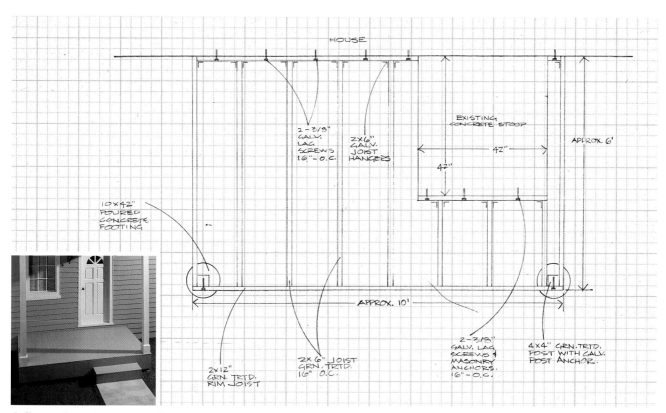

A floor plan

- HOUSE
- 2 - 3/8" GALV. LAG SCREWS 16"-O.C.
- 2x6" GALV. JOIST HANGERS
- EXISTING CONCRETE STOOP
- APPROX. 6'
- 42"
- 42"
- 10x42" POURED CONCRETE FOOTING
- APPROX. 10'
- 2x12" GRN. TRTD. RIM JOIST
- 2x6" JOIST GRN. TRTD. 16" O.C.
- 2 - 3/8" GALV. LAG SCREWS & MASONRY ANCHORS. 16"-O.C.
- 4x4" GRN. TRTD. POST WITH GALV. POST ANCHOR.

A floor plan shows overall project dimensions, the size and spacing of floor joists, the size and location of posts, and the types of hardware to be used for connecting and anchoring structural members.

Working Safely

Working on porches or patios means working outdoors. By taking common-sense precautions you can work just as safely outdoors as indoors, even though the exterior of your house presents a few additional safety considerations.

Building projects for porches and patios frequently require that you use ladders or scaffolding. Learn and follow basic safety rules for working at heights.

Any time you are working outside, the weather conditions should play a key role in just about every aspect of how you conduct your work: from the clothes you wear to the amount of work you decide to undertake. Plan your work days to avoid working in extreme heat. If you must work on hot days, take frequent breaks and drink plenty of fluids.

Wear sensible clothing and protective equipment when working outdoors, including: a cap to protect against direct sunlight, eye protection when working with tools or chemicals, a particle mask when sanding, work gloves, full-length pants, and a long-sleeved shirt. A tool organizer turns a 5-gallon bucket into a safe and convenient container for transporting tools.

Tips for Working Safely

• Work with a helper whenever you can. If you have to work alone, inform a friend or family member so he or she can check up on you periodically. If you own a portable phone, keep it handy at all times.

• Use cordless power tools when they will do the job—power cords are a frequent cause of worksite accidents. When using corded tools, plug them into a GFCI extension cord.

• Never work with tools if you have consumed alcohol or medication.

• Do not use power tools for tasks that require you to work overhead. Either find another way to access the task, or substitute hand tools.

Tips for Worksite Safety

Set up your worksite for quick disposal of waste materials. Use a wheelbarrow to transfer waste to a dumpster or trash can immediately. NOTE: Disposal of building materials is regulated in most areas—check with your local waste management department.

Create a storage surface for tools. Set a sheet of plywood on top of a pair of sawhorses to make a surface for keeping tools off the ground, where they are a safety hazard and are exposed to damage from moisture. A storage surface also makes it easy for you to locate a tool when needed.

Tips for Using Ladders

Do not stand on upper steps of stepladders, particularly when handling heavy loads.

Provide level, stable footing for extension ladders. Install sturdy blocking under ladder legs if the ground is uneven, soft, or slippery, and always drive a stake next to each ladder foot to keep the ladder from slipping away from the house.

Tools & Materials

You probably already own many of the basic hand and power tools needed to complete the projects shown in this book. Others, such as a flooring nailer, you may have to rent. For almost every project in this book, start with a tool kit consisting of the basic tools listed below, and add specialty tools as needed.

Whenever you build outdoor projects, like porches, use exterior-rated building materials and hardware whenever available. If you must use nonexterior-rated wood, like finish-grade pine, be sure to prime and paint it thoroughly.

Basic Hand Tools:
- tape measure
- pencil
- chalk line
- level
- wood chisel
- hammer
- circular saw
- drill and bits

Tools for porch construction include: flooring nailer (A), reciprocating saw (B), hammer drill with masonry bit (C), cordless drill (D), jig saw (E), carpenter's square (F), ratchet wrench (G), and speed square (H).

How to Use a Speed Square

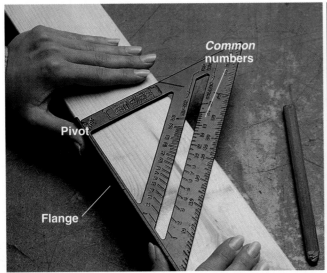

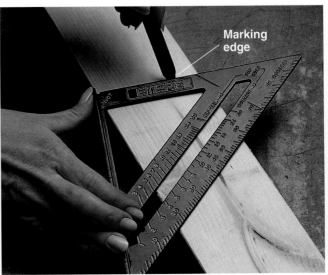

The speed square is a traditional roofer's tool that is very helpful for any projects that involve angle cutting. To use a speed square, you must either know or measure the slope of the line you want to mark, in inches per foot (page 110). Once you have the slope information, begin marking a cutting line onto a board by holding the flange of the speed square against the edge of the board. Look for the word *common* and the row of numbers aligned with it. Holding the end of the flange securely against the edge of the board, pivot the square so the *common* number equalling the rise of slope in inches per foot aligns with the same edge being pivoted against. Mark cutting line along the marking edge.

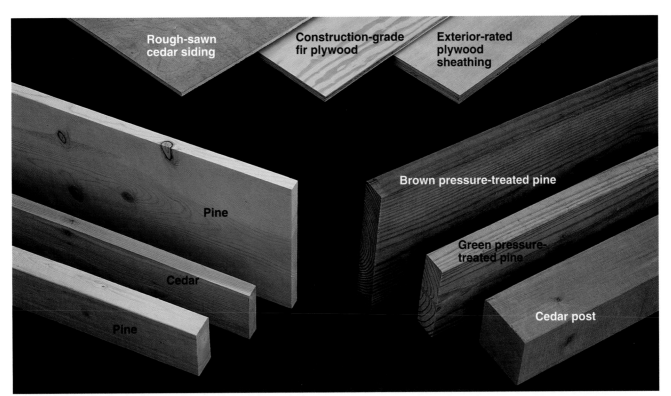

Labels on image: Rough-sawn cedar siding, Construction-grade fir plywood, Exterior-rated plywood sheathing, Pine, Cedar, Pine, Brown pressure-treated pine, Green pressure-treated pine, Cedar post

Lumber for porch and patio projects includes *sheet goods*, like ¼" cedar siding, ½" construction-grade plywood, and ¾" exterior-rated plywood sheathing. *Framing lumber* is usually 2×-dimensional lumber meant for strength, not appearance. It can be found in dimensions from 2 × 2 up to 2 × 12, and is available in cedar or green and brown pressure-treated pine. *Posts* can be cedar or treated pine. *Finish-grade lumber,* used for areas where appearance is important, is 1×-dimension pine or cedar.

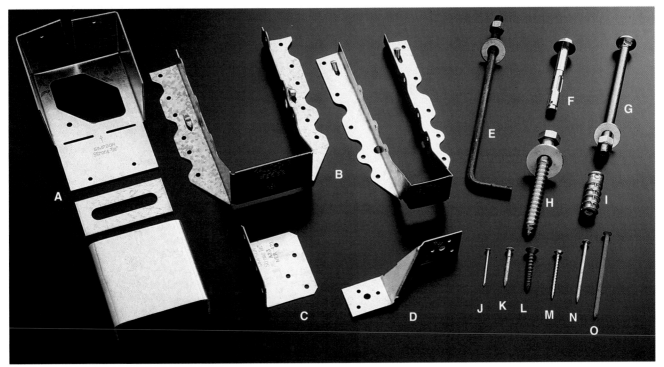

Labels on image: A, B, C, D, E, F, G, H, I, J, K, L, M, N, O

Connectors and fasteners include: post anchor with washer and pedestal (A), double- and single-joist hangers (B), angle bracket (C), rafter tie (D), J-bolt with nut and washer (E), masonry anchor bolt (F), carriage bolt with nut and washer (G), lag screw with washer (H), lead anchor sleeve (I), 4d galvanized nail (J), galvanized joist-hanger nail (K), self-tapping masonry screw (L), deck screw (M), 8d galvanized nail (N), and 16d galvanized nail (O).

Add a new front porch to your home, or replace an old, worn structure with a brand-new porch. A simple porch like the one shown above blends in well with rustic or unadorned house styles. Or, you can add railings, steps, and other decorative features to match the style of your house.

Porch-building Projects

Whether it is located in the front, back, or side of your house, a porch is a prominent feature that should be built or updated only in accordance with a detailed plan. Working with a design professional, use your design ideas to create a plan for a new or updated porch that is attractive, functional, and within your budget.

The intent of the projects on the following pages is to demonstrate how to turn your building plan into a finished product. Do not attempt to create these projects without an approved plan of your own. In addition to a plan, you will need a building permit from your local building department for most porch projects.

The best porch designs reflect the style of the house, while meeting your needs for use. If you plan on spending many evenings relaxing on the porch, you will want a larger structure than if your main goal is provide shelter at the entry to your home. The same standards hold true for updating projects, like adding a railing to an open porch, or screening in a porch.

This sections shows:
• Building Porches (pages 120 to 147)
• Building & Installing Porch Railings (pages 148 to 151)
• Building Wood Porch Steps (pages 152 to 157)

Options for Building & Updating Porches

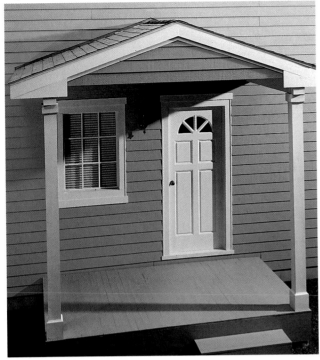

An open front porch is the simplest type of porch, but the deck can be no more than 20" above the ground in most areas. Because there are no railings, an open porch is a good choice for homeowners who like to visit with neighbors and passersby. Pages 120 to 147.

Porch railings are a necessary safety feature on porches that are 20" or more above the ground, but they are also an attractive design element that can be integrated into your new porch design, or added to your existing porch. Pages 148 to 151.

Wood porch steps are fairly easy to construct. Though not as maintenance-free as concrete steps, they better match the style of most porches. Step railings can be designed and built to match the porch railing. Pages 152 to 157.

Screened-in porches are very popular conversion projects for homeowners in areas where insects are a problem. The screening-in process is surprisingly simple.

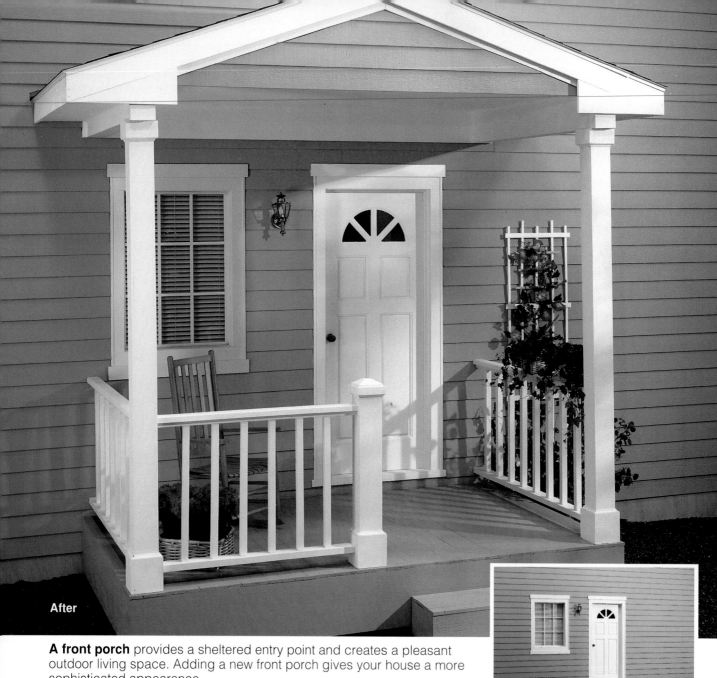

After

Before

A front porch provides a sheltered entry point and creates a pleasant outdoor living space. Adding a new front porch gives your house a more sophisticated appearance.

Building Porches

Adding a front porch to your house is a major project. But with thorough preparation and a detailed construction plan, a successful porch-building project can be accomplished by most do-it-yourselfers.

A porch is a permanent part of your home, so make sure the foundation and structure are sturdy and meet all local building codes. Also pay close attention to design issues so the size and style of the porch make sense with the rest of your house. Use the techniques illustrated in this project as a guide to help you convert your own porch plan into a reality.

This project shows:

- How to Install Ledger Boards & Posts (pages 124 to 128)
- How to Install Deck Joists (pages 129 to 130)
- How to Install Porch Floors (pages 131 to 133)
- How to Install Beams & Trusses (pages 134 to 137)
- How to Install Roof Coverings (pages 138 to 140)
- How to Wrap Posts & Beams (pages 141 to 142)
- How to Finish the Cornice & Gable (pages 143 to 144)
- How to Install Soffits & Ceilings (pages 145 to 146)
- Tips for Applying Finishing Touches (page 147)

Anatomy of a Porch

The basic parts of a porch include the roof, the posts and beams, the floor and floor deck, the support system (ledger board and post footings), trim, and optional elements like railings and steps. See pages 122 to 123 for more information on these systems.

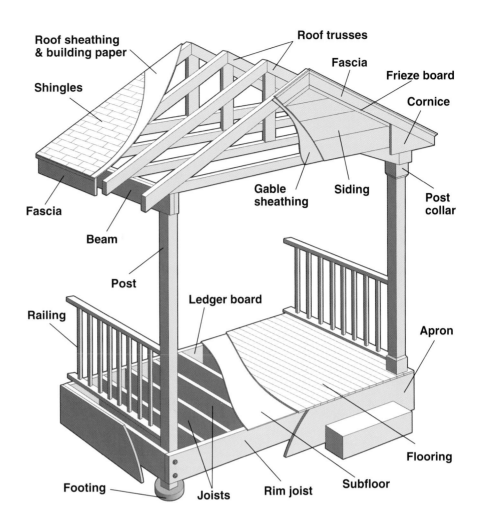

Roof sheathing & building paper

Roof trusses

Fascia

Frieze board

Cornice

Shingles

Gable sheathing

Siding

Post collar

Fascia

Beam

Post

Ledger board

Railing

Apron

Footing

Joists

Rim joist

Subfloor

Flooring

Tips for Building Front Porches

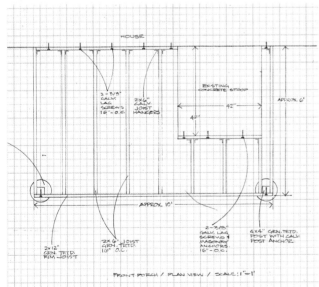

Work from a construction plan. Because a porch is a permanent part of your house, you will need a building permit and a detailed construction plan before you start (pages 112 to 113). The plan should include a floor plan, like the one shown above, and an elevation drawing. For your convenience, also create a comprehensive materials list and an assembly plan.

Rafter chords   Rafter tail

Bottom chord

Use pre-built roof trusses for porch construction. They are easier to work with than site-built rafters. When ordering trusses, you must know the roof pitch (page 110), the distance to be spanned, and the amount of overhang past the beams. Trusses can be purchased in stock sizes, or custom-ordered at most building centers—consult with the salesperson to make sure you get the right trusses for your project.

Elements of a Porch: The Foundation

Concrete footings, cast in tubular forms, support the porch posts. Post anchors, held in place with J-bolts, secure the posts to the footings. Frost footings, required for porches, should be deep enough to extend below the frost line, where they are immune to shifting caused by freezing and thawing.

Existing concrete steps with sturdy footings can be used to support the porch deck—an easier option than removing the concrete steps. Excavate around the steps to make sure there is a footing, and that it is in good condition.

Elements of a Porch: The Deck & Floor

Joist hanger

Ledger

Joist

Rim joist

A series of parallel joists supports the floor. Joist hangers anchor the joists to the ledger and the front rim joist, which acts like a beam to help support both the joists and the floor. Local building codes may contain specific requirements for thickness of the lumber used to make the rim joist. The face of the rim joist often is covered with finish-grade lumber, called an "apron," for a cleaner appearance.

Tongue-and-groove porch boards installed over a plywood subfloor is common porch-floor construction. The subfloor is attached to the joists. The porch boards, usually made of fir, are nailed to the subfloor with a floor nailer tool (page 116).

Elements of a Porch: The Posts & Beams

The posts and beams support the weight of the porch. Posts are secured by post anchors at the footings. Beams are connected to the posts with post saddles, and attached to a ledger board at the house using double-joist hangers. For most porch projects, doubled 2 × 6 or 2 × 8 can be used to build beams.

Wraps for posts and beams give the rough framing lumber a smoother, more finished appearance, and also make them look more substantial and in proportion with the rest of the porch. Use finish-grade 1 × 6 and 1 × 4 pine to wrap 4 × 4 posts.

Elements of a Porch: The Roof

The porch roof is supported and given its shape by sloped rafters or trusses. Prebuilt trusses, like those shown above, are increasingly popular among do-it-yourselfers. The ends of the trusses or rafters extend past the porch beams to create an overhang that often is treated with a soffit (pages 145 to 146). The trusses or rafters provide support and a nailing sur-

face for the roof sheathing, a layer of building paper, and the roofing materials—like asphalt shingles. The area of the porch below the peak of the roof, called the "gable," usually is covered with plywood sheathing and siding. Metal flashing is installed between the roof and the house to keep water out of the walls.

1 Lay out the location for the porch ledger board, based on your project plans. Mark the center of the project area onto the wall of your house to use as a reference point for establishing the layout. Measure out from the center and mark the endpoints of the ledger location. Make another mark ½" outside the endpoint to mark cutting lines for siding removal. According to most codes, the siding must be removed before the ledger board is installed.

2 Mark the ledger height at the centerline—if you are building over old steps, the top of the ledger should be even with the back edge of the steps. Use a straightedge and a level to extend the height mark out to the endpoints of the ledger location. Mark cutting lines for the ledger board cutout on the siding, ½" above the ledger location, then measure down from the top cutting line a distance equal to the width of the ledger board plus 1". Mark the bottom of the cutout area at that point, and extend the mark across the project area with the level and straightedge.

How to Install Ledger Boards & Posts

Ledger boards and posts support the roof and the deck of a porch. A ledger board is a sturdy piece of lumber, usually a 2 × 6 or 2 × 8, that is secured to the wall of a house to support joists or rafters for the porch. The posts used in most porch projects are 4 × 4 or 6 × 6 lumber that is attached to concrete footings with post-anchor hardware. Proper installation of the posts and ledgers is vital to the strength of the porch.

In most cases, porches are built with posts at the front only. A ledger is installed at deck level to support the floor, and another is sometimes installed at ceiling level to anchor the beams and the rafters or trusses.

If you are building your porch over an old set of concrete steps (page 122), make a cutout in the deck-level ledger board that is the same width and position as the steps, and attach the cut section to the top of the top step with masonry anchors.

Ledgers must be attached to the wall at framing member locations, or attached directly to the house rim joist (page 111) if the rim joist is at the correct height. Find and mark the framing member locations before starting ledger installation.

Everything You Need:

Tools: basic hand tools, caulk gun, framing square, mason's string, straightedge, plumb bob.

Materials: construction plans, framing lumber, drip edge flashing, caulk, concrete, tubular form, post anchor, joist-hanger nails, lag screws.

3 Remove the siding at the cutting lines (page 111). For wood siding, set the blade of a circular saw so it cuts to thickness of the siding, and cut along the cutting lines. Finish the cuts at the corners with a wood chisel. Remove the siding. You do not generally need to remove the wall sheathing between the siding and the framing members.

4 Cut a piece of metal or vinyl drip-edge flashing to fit the length of the cutout area. Apply caulk or exterior panel adhesive to the back face of the flashing—do not use fasteners to attach it. Slip the flashing behind the siding at the top of the cutout.

5 Cut a ledger board to the size specified in your construction plan. Because the end of the outer deck joist at each side of the project will butt against the wall sheathing in most cases, ledgers should be cut 4" shorter than the full planned width of the porch to create a 2" gap for the joist at each end. Also cut the rim joist for the porch (usually from 2 × 12 lumber) according to your project plans. Lay the ledger board next to the rim joist to gang-mark deck joist locations onto the ledger and the rim joist. To allow for the difference in length between the ledger and the rim joist, set a 2 × 6 spacer at each end of the ledger. Mark the deck joist locations onto the ledger and the rim joist according to your construction plan. In the project above, we gang-marked deck joist locations 16" apart on center, starting 15¼" in from one end of the rim joist.

(continued next page)

Ledger-section location

Ledger-section location

OPTION: If you are attaching a section of the ledger to concrete steps (page 122), set the ledger in position on the back of the steps, and mark cutting lines onto the full-length ledger board at the edges of the steps. Cut at the cutting lines to divide the ledger into three sections.

6 Position the ledger board in the cutout area, up against the drip-edge flashing. Tack in place with duplex nails. Drill two counterbored pilot holes into the ledger at framing member locations, or at 16" intervals if attaching at the rim joist. Drive ⅜ × 4" lag screws, with washers, into the pilot holes to secure the ledger. Install all ledger sections that attach to the wall.

3 ft.

5 ft.

4 ft.

Front of project area

Mason's string

Project edge

Batter board

7 Establish square lines for the sides of the porch. First, build 3-piece 2 × 4 frames, called "batter boards," and drive one into the ground at each side, 12" past the front of the project area, aligned with the project edge. Drive a nail at each end of the ledger, and tie a mason's string to each nail. Tie the other end of each string to a batter board. Square the string with the ledger, using the 3-4-5 method: mark the ledger board 3 ft. from the end, then mark the mason's string 4 ft. out from the same point. Adjust the mason's string until the endpoints from the 3-ft. and 4-ft. marks are exactly 5 ft. apart, then retie.

8 Mark locations for the centers of the porch posts onto the mason's strings by measuring out from the ledger board, using your construction plan as a guide. Use a piece of tape to mark the mason's string. Make sure the mason's string is taut.

9 Transfer the location for the post centers to the ground by hanging a plumb bob from the post marks on the mason's string. Drive a stake at the post-center location, then set an 8"-diameter tubular concrete form onto the ground, centered around the stake. Mark the edges of the form onto the ground, then remove the form and the stake, and dig a hole for the form past the frost-line depth (page 16). Avoid moving the mason's string. Set the tubular form into the hole so the top is about 2" above ground. Use a level to make sure the top of the form is horizontal. Excavate for and install both tubular forms.

(continued next page)

10 Fill the tubular forms with fresh concrete. Smooth the surface of the concrete with a trowel or float, then insert a J-bolt into the fresh concrete. Use a plumb bob to find the point on the surface of the concrete that is directly below the tape mark on each mason's string. Insert a J-bolt into the concrete at that point. The threaded end of the J-bolt should extend up at least 2". Let concrete cure for three days.

11 Set a metal post anchor over the J-bolt, and secure with a washer and nut. NOTE: Some post anchors have a pedestal that fits over the J-bolt to support the post. Cut a post that is at least 6" longer than the planned post height. With a helper, set the post into the post anchor and secure with 10d galvanized nails driven through the pilot holes in the post anchor. Install both posts.

12 Brace each post with a pair of 2 × 4s attached to stakes driven in line with the edges of the posts, but outside of the project area. Secure the posts to the stakes with deck screws, then use a level to make sure the posts are plumb. Attach the braces to the posts with deck screws.

Techniques for
Building Porches

How to Install
Deck Joists

With the support for the porch project in place, begin installing the deck joists. Joists usually can be made from 2 × 6 lumber, but check your local codes to be sure. Install them at the edges of the project, and at 16" intervals, perpendicular to the house. Also install a 2 × 12 porch rim joist between the front posts of the porch. For drainage, the deck joists should slope away from the house at a rate of ⅛" per foot.

Everything You Need:

Tools: basic hand tools, ratchet & socket set.

Materials: framing lumber, lag screws, joist hangers, corner brackets, masonry anchoring hardware.

Level line
Slope line (⅛" per foot)

1 With post braces still in place, run a mason's string between a post and the end of the ledger. Use a line level to make sure the string is level, then measure down ⅛" for every foot of distance between the ledger and the post to establish a slope line. Mark a slope line on each post.

2 Cut outer joists to fit between the back of the ledger and the front of each post, using the angle created by the slope line and the post as a guide for cutting the ends of the joists. Attach the outer joists to the ends of the ledger and the posts with deck screws—you may need to bend up the drip-edge flashing above the ledger.

Ledger
Step ledger (uninstalled)

3 Attach joist hangers to the ledger and to the rim joist at the joist locations (step 5, page 125) with galvanized joist-hanger nails. TIP: Slip a 2 × 6 scrap into each hanger before nailing to make sure the hanger holds its shape during installation.

4 Tack the porch rim joist in position: the top of the rim joist should be flush with the tops of the outer deck joists. Drill four counterbored pilot holes through the rim joist and into each post, then drive ⅜ × 4" lag screws with washers at each pilot hole to secure the rim joist.

(continued next page)

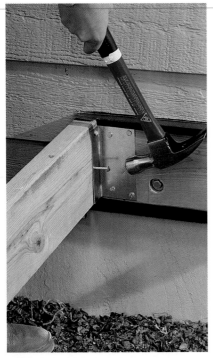

5 Install metal corner brackets at each of the four inside corners, to stabilize the frame. Use joist-hanger nails to fasten the corner brackets.

6 Cut the remaining deck joists that fit all the way from the ledger to the porch rim joist—cut the same end angles that were cut for the outer joists (step 2, page 129). Install the deck joists in the joist hangers with joist-hanger nails.

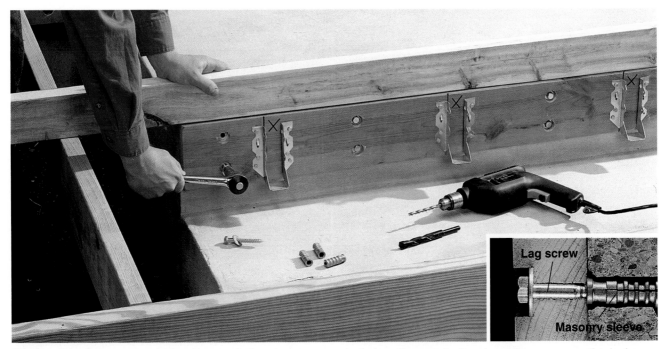

OPTION: If building over old steps, attach the step ledger board to the riser of the top step, using masonry anchors (inset). Lay a straightedge across the joists next to the steps as a reference for aligning the top of the ledger board. Shim under the ledger to hold it in position, then drill counterbored, ⅜"-diameter pilot holes for lag screws into the ledger. Mark the pilot hole locations onto the riser, then remove the ledger. Drill holes for ⅜"-diameter masonry-anchor sleeves into the riser with a hammer drill and masonry bit. Drive the sleeves into the holes with a maul or hammer, then attach the step ledger with ⅜ × 4" lag screws driven into the masonry sleeves. Drive pairs of lag screws at 16" intervals.

Techniques for Building Porches

How to Install Porch Floors

Like most floors inside your house, a porch floor is made of a subfloor, usually plywood, topped with porch boards. The subfloor provides a stable base over the joists, which usually run parallel to the house for maximum strength. Porch boards are nailed directly to the subfloor—we rented a floor nailer for this task (page 116).

Everything You Need:

Tools: basic hand tools, caulk gun, speed square, jig saw, floor nailer, mallet, nail set, hand saw.

Materials: plywood, tongue-and-groove flooring, finish flooring, floor nailer nails, deck screws.

1 Begin laying the plywood subfloor—we used ¾"-thick exterior plywood. Measure and cut plywood so any seams fall over deck joists, keeping a slight (⅛") expansion gap between pieces. Fasten plywood pieces with 1½" deck screws. If installing the floor over old steps, apply exterior-grade construction adhesive to the steps to bond with the plywood.

2 Notch plywood to fit around posts. Also nail a 2 × 4 nailing cleat to the edges of the post that are not fitted against joists. Make sure the cleat is level with the tops of the joists.

3 Cut a starter board from a tongue-and-groove porch board by ripcutting the grooved edge off of the board with a circular saw. Cut the starter board 1" longer than finished length, including a ¾" overhang for the porch apron.

4 Set the starter board next to the post, with the tongue edge pressed against the post. Mark the location of the post onto the porch board, measure and mark the cutting depth to fit around the post, then notch the board with a jig saw.

(continued next page)

131

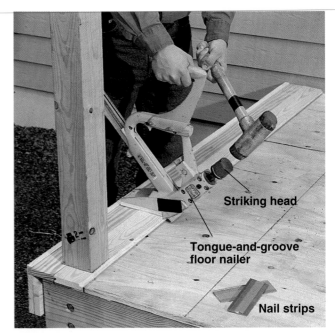

5 Make a cleat and spacer from scrap lumber the same thickness (usually ¾") as the skirt board you will install at the side of the porch. Sandwich the cleat and spacer together and attach them to the outer joist so the cleat is on the outside and at least 2" above the top of the joist. The spacer should be below the top of the joist. The cleat provides a secure, straight edge for aligning the first porch board, and the spacer creates an overhang for the skirt board.

6 Butt the notched porch board (step 4, previous page) against the cleat so it fits around the post, and nail in place. If using a tongue-and-groove floor nailer, load a nail strip, then position the nailer over the exposed tongue and rap the striking head with a mallet to discharge and set the nails. Nail at 6" to 8" intervals, then cut and position the next porch board (notched for the post, if needed) so the groove fits around the tongue of the first board, and nail in place.

7 Continue installing porch boards. Draw reference lines on the subfloor, perpendicular to the house, to check the alignment of the porch boards. Measure from a porch board to the nearest reference line occasionally, making sure the distance is equal at all points. Adjust the position of the next board, if needed.

8 Notch porch boards to fit around the other front post before you install them, then ripcut the last board to fit (create an overhang equal to the starter-board overhang). Position the last board, and drive galvanized finish nails through the face of the board and into the subfloor. Set the nail heads with a nail set.

9 Trim the exposed porch board ends so they are even. First, mark several porch boards ¾" out from the front edge of the rim joist to create an overhang that will cover the top of the apron (see next step). Snap a chalk line to connect the marks, creating a cutting line. Use a straightedge as a guide and trim off the boards at the cutting line with a circular saw. Use a hand saw to finish the cuts around the posts.

10 Cut aprons from exterior plywood to conceal the outer joists and the rim joist. Cut the side aprons (we used ¾"-thick plywood) so they fit flush with the front edges of the posts, then install them beneath the porch board overhangs and nail in place with 8d siding nails. After the side aprons are in place, cut the front apron long enough to cover the edge grain on the side aprons, and nail it to the rim joist.

OPTION: Build a wooden step and cover the surface with porch boards. The step above was built with a framework of 2 × 8s (toenailed to the front skirt), wrapped with exterior plywood, then covered with porch boards. See pages 152 to 157 for more information on building porch steps.

133

How to Install Beams & Trusses

Beams and trusses or rafters support the porch roof. Installing them takes patience and care. Use prebuilt trusses to simplify the project (page 123). Always have at least one helper on hand when you install trusses: they are awkward to handle. For beams, we used doubled 2 × 8s.

For information on installing the roof ledger, see pages 124 to 126.

Everything You Need:

Tools: basic hand tools, speed square, stud finder, torpedo level, hand saw, clamps, straightedge.

Materials: ladders, framing lumber, prebuilt roof trusses, double-joist hanger, nails.

1 Measure to find the midpoint of the porch deck and mark it on the house, then transfer the centerline to the peak area of the planned roof, using a straight 2 × 4 and a carpenter's level. Refer to your construction plan, then measure up from the porch deck and mark the top and bottom of the roof ledger onto the siding, near the center mark. Also mark the ledger height at the ends of the project area, then connect the height marks with a chalk line.

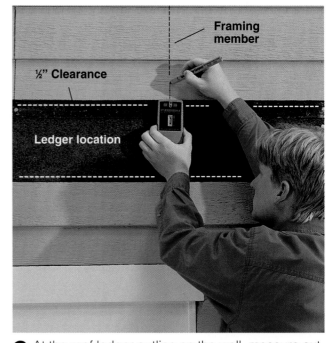

2 Mark a 2 × 6 to cut for the roof ledger by setting it on the deck so it extends past the edges of both front posts. Mark the outside edges of the posts onto the 2 × 6, then make another mark directly above each edge of the deck. Cut the ledger to length, and make a reference mark at the midpoint of the board.

3 At the roof ledger outline on the wall, measure out in each direction from the centerpoint and finish the outline so it is the same length as the ledger board. Enlarge the outline by ½" on all sides, then remove the siding (page 111) in the outlined area. Use an electronic stud finder to locate framing members.

4 Attach the roof ledger with pairs of ⅜ × 4" lag screws driven at framing member locations (page 126). Lay out the locations for the beams onto the ledger, according to your construction plan. Insert a pair of 2 × 8 scraps into the double-joist hanger to help it keep its shape when you nail it. Position the hanger against the ledger, using a torpedo level to make sure it is plumb. Fasten the double-joist hangers to the ledger with joist-hanger nails.

5 Mark the front posts at the height of the bottoms of the double-joist hangers. To make sure the marks are level, set a straight board in each joist hanger and hold the free end against the post. Use a carpenter's level to adjust the height of the board until it is level, then mark the post where it meets the bottom of the straight board. Draw cutting lines on all sides of the post at the height mark.

6 Steady the post, and trim off the top at the cutting line. SAFETY TIP: Have a helper brace the post from below, but be careful not to drop the cutoff post end in the area.

7 Make the beams (we used pairs of 2 × 8s, cut to length, then nailed together with 16d common nails). The beams should extend from the double-joist hangers, past the fronts of the posts (1½" in our project). When nailing boards together to make beams, space nails in rows of three, every 12" to 16". For extra holding power, drive the nails at a slight angle.

(continued next page)

How to Install Beams & Trusses (continued)

8 Lay out truss locations onto the tops of the beams, starting at the beam ends that will fit into the joist hangers. Mark both edges of each truss, drawing an "X" between the lines for reference. Generally, trusses should be spaced at 24" intervals—check your construction plan for exact placement.

9 Set a metal post saddle onto the top of each front post, and nail in place with joist-hanger nails. With a helper, raise the beams and set them into the post saddles and double-joist hangers. Secure the beams in the double-joist hangers with joist-hanger nails.

10 If your beams are thinner than your posts (as above) cut plywood spacers and install them between the inside edges of the beams and the inner flange of the post saddle. The spacers should fit snugly, and be trimmed to roughly the size of the saddle flanges. Drive joist-hanger nails through the caps, into the spacers and beams.

11 With help, hoist the first truss into position. Turn trusses upside down to make them easier to handle when raising. Rest one end of the truss on a beam, then slide the other end up and onto the opposite beam. Invert the truss so the peak points upward, and position it against the house, with the peak aligned on the project centerline.

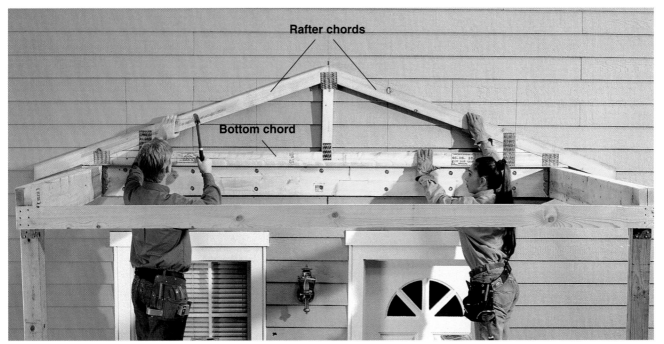

12 Make sure the first truss is flush against the siding, with the peak aligned on the project centerline. Nail the rafter chords and bottom chord of the truss to the house at framing member locations, using 20d common nails. Lift the remaining trusses onto the beams. TIP: If you are installing trusses with a pitch of 4-in-12 or less, you may find it easier to lift all trusses onto the beams before continuing with installation. For steeper pitches, lift and nail trusses one at a time.

13 Install the rest of the trusses at the locations marked on the beams, working away from the house, by toenailing through the bottom chords and into the beams with 8d nails. Nail the last truss flush with the ends of the beams. NOTE: If the bottom chord of the first truss overhangs the beams, install the rest of the trusses with equal overhangs.

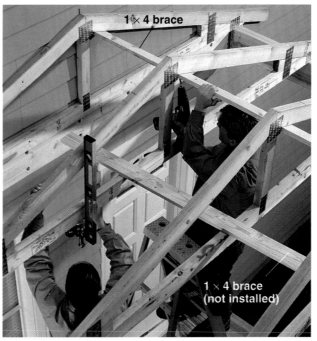

14 Attach 1 × 4 braces to the underside of each row of rafter chords. Use a level to plumb each truss before fastening it to the braces with 2" deck screws.

Nailing strip

How to Install Roof Coverings

The roof covering for a porch includes plywood roof sheathing, building paper, shingles, fascia boards, and other trim pieces. If you have never installed roof coverings before, get additional information on roofing techniques.

Everything You Need:

Tools: basic hand tools, ladder, hand saw, reciprocating saw, aviator snips.

Materials: plywood sheathing, finish-grade lumber, 30# building paper, metal roll flashing, 3-tab shingles, roofing nails, drip-edge flashing, roof cement.

1 Cut and attach 2 × 4 nailing strips to the rafter chords of the front truss, using 2½" deck screws. Nailing strips create a nailing surface for the roof sheathing overhang. Cut them to the same dimensions as the rafter chords.

Rafter tail

2 Cut ¾" exterior-grade plywood sheathing to cover the trusses and nailing strips. The sheathing should be flush with the ends of the rafter tails. Cut sheathing pieces so seams fall over rafter locations, and install them with 8d siding nails or deck screws.

3 Fill in the rest of the sheathing, saving the pieces that butt together at the peak for last. Leave a ¼" gap at the peak.

Gable sheathing

4 Cut ½"-thick plywood to cover the gable end of the roof. Measure the triangular shape of the gable end, from the bottom of the truss to the bottoms of the nailing strips. Divide the area into two equal-sized triangular areas, and cut plywood to fit. Butt the pieces together directly under the peak, and attach them to the front truss with 1½" deck screws.

Fascia board

Frieze board

Fascia board   Gable sheathing

Frieze board   Truss

5 Cut 1 × 4 frieze boards to fit against the plywood gable sheathing, beneath the nailing strips. Attach the frieze boards to the gable sheathing with 1¼" deck screws. Then, cut fascia boards long enough to extend several inches past the ends of the rafter tails. Nail the fascia boards to the nailing strips, with the tops flush with the tops of the roof sheathing.

Gable-end fascia board

Side fascia board (plowed)

6 Measure for side fascia boards that fit between the house and the back faces of the gable-end fascia boards. Cut the fascia boards to fit, then attach with galvanized 8d finish nails driven into the ends of the rafter tails. Make sure the tops of the fascia boards do not protrude above the plane of the roof sheathing. NOTE: If you plan to install soffits, use fascia boards with a plowed groove for soffit panels.

7 Trim off the ends of the gable-end fascia boards so they are flush with the side fascia boards, using a handsaw. Drive two or three 8d finish nails through the gable-end fascia boards and into the ends of the side fascia boards.

(continued next page)

8 Remove siding above the roof sheathing to create a recess for metal roof flashing. Make a cut about 2" above the sheathing, using a circular saw with the cutting depth set to the siding thickness. Then make a cut flush with the top of the roof, using a reciprocating saw held at a very low angle. Connect the cuts at the ends with a wood chisel and remove the siding. See page 111 for more information on removing siding.

9 Install building paper, drip-edge flashing, and shingles as you would for a standard roofing project (see *Exterior Repairs & Projects*). Slip pieces of metal step flashing behind the siding above the cutout area as you finish rows of shingles, sealing the seams with roof cement.

10 Finish shingling and flashing the roof. Make sure shingle tabs are staggered in regular patterns, with a consistent exposed area on shingle tabs. Cut off shingle tabs and use them to create the roof ridge.

How to Wrap
Posts & Beams

Even in the simplest front porch, standard posts and beams can look spindly and plain. Rather than spending the money for large timbers, it is a common practice among builders to wrap the posts and beams with finish-grade lumber to give them a more proportional look. We used finish-grade pine to give the posts in our porch project the appearance of solid 6 × 6 stock, and to conceal the fact that our beams are actually made from doubled 2 × 8s.

Everything You Need:

Tools: basic hand tools.

Materials: finish lumber, siding nails, plywood strips.

Inner beam wrap board

Plywood spacers

1 Cut wrap boards for the inner sides of the beams to the same length as the beams, using finish lumber wide enough to cover the beams and any metal saddles or joiners. We used 1 × 10, but sanded ¾" plywood can be used instead. Attach the inner side boards to the beams with 8d siding nails—in the project above, we added ½" plywood strips at the top and bottom of the beam to compensate for the ½" spacers in the metal post saddles.

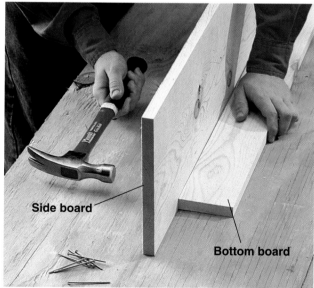

Side board

Bottom board

2 Cut strips of wood to cover the bottoms of the beams. Position each strip next to a board cut the same size as the inner beam wrap. The difference in length between the side board and the bottom board should equal the distance of the beam overhang at the post. Preassemble the bottom board and side board by driving 8d finish nails at the butt joint, making sure to keep the joint square. Attach the assembly to the beam so the free end of the bottom board forms a butt joint with the inner beam wrap board.

3 Cut boards to create an end cap for each beam—we cut a piece of 1 × 10 to fit over the ends of the beam and the beam wrap, and attached it to a piece of 1 × 4 cut to cover the gap beneath the beam overhang. Nail end caps over the end of each beam.

(continued next page)

4 Cut boards for wrapping the posts so they span from the floor to the bottoms of the wrapped beams. For a 4 × 4 post, two 1 × 4s and two 1 × 6s per post can be used. Nail a 1 × 6 to the front of the post, overhanging ¾" on the outside edge. Nail a 1 × 4 to the outer face of the post, butted against the 1 × 6.

5 Preassemble the other two wrap boards, nailing through the face of the 1 × 6 and into the edge of the 1 × 4. Set the assembly around the post, nailing the 1 × 6 to the post and nailing through the other 1 × 6 and into the edge of the 1 × 4 (there will be a slight gap between the second 1 × 4 and the post).

Furring strip

6 Cut pieces of finish lumber to fit around the bases of the posts (called "post collars"). We used 1 × 6 to create the bottom post collars, and 1 × 4 to create the top collars where the posts meet the beams. Nail the collars together with 4d finish nails. TIP: Cut pieces so the front collar board covers the end grain of the side boards.

7 Roof ledgers often are visible after the porch ceiling is installed, so cover the ledger with finish lumber. If the ledger protrudes past the siding, cut a furring strip to cover the gap between the inside face of the ledger cover and the siding. Cut the ledger cover and furring strips to fit, and install with 8d nails. If the ledger extends past the outer face of the beam, the easiest solution is to paint it to match the siding.

How to Finish the Cornice & Gable

The gable and the cornice are prominent features on the front of any porch. The gable is the area just below the peak, which is usually covered with trim and siding material. TIP: Paint siding materials before installation. The cornice, sometimes called the "cornice return" or the "fascia return," is usually fitted with trim that squares off the corner where it meets the soffit.

Everything You Need:

Tools: basic hand tools, miter box, straightedge guide.

Materials: plywood, framing lumber, finish-grade lumber, cove molding, nails, caulk.

Gable trim

Cornice

The cornice and gable are finished to match the siding and trim on your house. Use plywood or finish-grade lumber to make the cornice, and use siding that matches your house for the gable trim. Caulk seams at the peak of the gable, and between the fascia boards and the cornice (inset).

How to Install a Cornice

1 At each end of the front porch, measure the area from the end of the gable fascia to a spot about 6" inside the porch beam. Lay out a triangular piece of plywood or finish-grade lumber to fit the area, using a carpenter's square to create right angles. Cut out the cornice pieces, using a circular saw and straightedge.

2 Test-fit the cornice pieces over the ends of the porch gable, then install with 8d finish nails driven into the ends of the beams, and 4d nails driven up through the ends of the cornice pieces and into the underside of the gable fascia. Use a nail set to embed the heads of the nails below the surface of the wood, being careful not to split the cornice pieces.

Gable fascia

Frieze board

1 Measure the dimensions of the area covered by the gable sheathing (page 139) on the house. If you have installed fascia and frieze boards, measure from the bottom of the frieze boards. Add 2" of depth to the area to make sure that siding will cover the edge of the ceiling once the ceiling and soffits are installed (pages 145 to 146). Snap a horizontal chalk line near the bottom of the gable sheathing to use as a reference line for installing the siding.

2 Mark a cutting line that matches the slope of the roof onto the end of one piece of siding. Use a framing square or a speed square (page 116) to mark the slope line. OPTION: Position a scrap board on the horizontal chalk line on the gable sheathing, and mark the points where the edges of the board intersect with the frieze board: connect the points to establish the slope line. Cut the siding or scrap board on the slope line and use it as a template to mark siding for cutting. Cut the bottom siding board to length.

3 Use 4d siding nails to install the bottom siding board so it is flush with the bottom edges of the frieze boards—the bottom edge of the siding board should be 2" lower than the bottom of the gable sheathing. Cut the next siding board so it overlaps the first board from above, creating the same amount of exposed siding as in the rest of the house. Be careful to keep the siding level. Continue cutting and installing siding pieces until you reach the peak of the gable.

Techniques for
Building Porches

How to Install Soffits & Ceilings

Soffits are panels that close off the area between rafter ends and the side of the porch. They can be nailed directly to the rafters, or attached to nailers on the porch beams so they are horizontal. A porch ceiling can be made from plywood or tongue-and-groove boards to gives the porch a finished appearance.

Everything You Need:

Tools: basic hand tools, straightedge guide, torpedo level, caulk gun, miter box.

Materials: plywood, plowed fascia board, nails, nailer.

Plowed groove

Option: Attach soffit directly to rafters. Attach a fascia board with a plowed groove (inset) to the rafter ends (page 139, step 6). Measure from the back of the fascia plow groove to the beam, following the bottom of the rafter tail, to establish the required width for the soffit panel. Measure the length from the house to the cornice or gable, then cut a piece of ⅜"-thick plywood to these dimensions. Insert one edge of the plywood panel into the plowed groove, and press the soffit panel up against the rafter tail. Nail the panel in place with 4d galvanized common nails, then caulk the edges before painting the panel.

How to Install Horizontal Soffit Panels

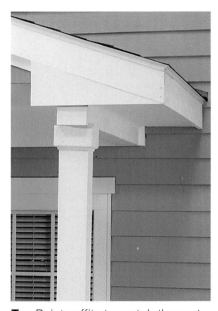

1 Install a fascia board with a plowed groove (page 139, step 6). Use a torpedo level to transfer the top height of the groove to the beam, near one end. Mark the groove height at the other end of the beam, then connect with a chalk line. Install a 2 × 2 nailer just above the chalk line.

2 Measure from the back of the plowed groove to the beam, just below the nailer, to find the required width of the soffit panel. Measure the length, then cut a piece of ⅜"-thick plywood to fit. Insert one edge of the soffit panel into the plow, then nail the other edge to the nailer with 4d nails.

TIP: Paint soffits to match the rest of the porch trim. Add quarter-round molding at the joint between the soffit and the beam, or fill the gaps with tinted exterior-grade caulk.

How to Install a Porch Ceiling

1 To create nailing surfaces for ceiling materials, cut 2 × 4s to fit between rafters, spaced 24" on center, and flush with the bottoms of the rafters. NOTE: If you plan to install ceiling lights, have them wired before proceeding.

2 Measure the ceiling space and cut ceiling materials to fit (we used 4 × 8 sheets of ⅜"-thick plywood). Cut ceiling pieces so the seams fall on the centers of the rafters and nailing strips. Use 4d galvanized nails to attach the plywood to the rafters and nailers. Space nails at 8" to 12" intervals. Do not drive nails next to one another on opposite sides of joints.

3 Install molding around the edges of the ceiling to cover the gaps and create a more decorative look. Simple ¾" cove molding, which does not require any complicated coping cuts or installation techniques, is used for the above project. Miter the corners and attach with 4d finish nails. Set the nail heads slightly, then cover with wood putty before painting.

Tips for Applying Finishing Touches

A few final touches add personality and a completed look to your porch-building project. There are a wide range of decorative trim types that can be purchased or built, then installed to dress up the porch or help it blend with the style of your house. Look through millwork catalogs to find ideas for trimming out your new porch. Also visit salvage yards to find authentic millwork, like gingerbread types, that will blend in with the rest of your house.

Avoid getting carried away with decorative trim. A few elegant touches will go a long way.

Use gable ornamentation to soften the hard lines of the gable peak. Fan-style trim (above) and many other trim types that fit into a peak are made to fit a range of different peak angles. Make sure to measure your peak angle carefully before purchasing or ordering gable ornamentation.

Install decorative trim to enhance the appeal of your porch. The scalloped fascia boards above were cut from plain 1 × 6 pine, using a jig saw, to add a touch of flair to a plain porch gable.

Put stock moldings, sold at any building center, to creative use on a porch. The simple cove molding above is installed at the joint between a post collar and the post, to create more graceful lines.

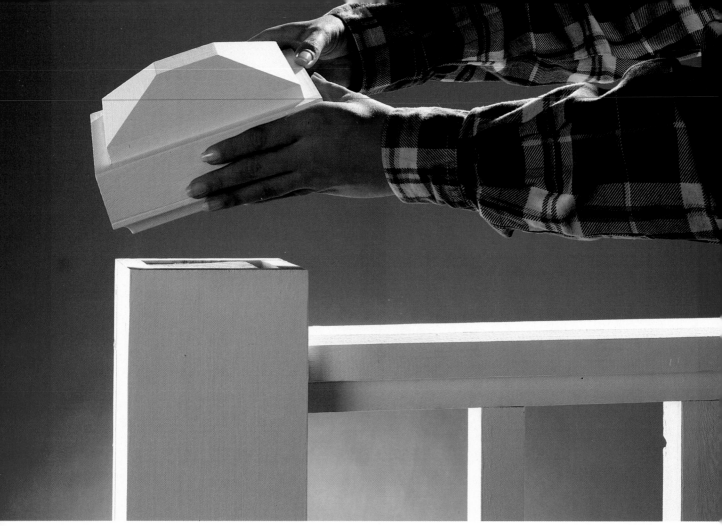

Add preformed post caps to posts that support porch railings. Post caps are sold in a wide range of styles and sizes at most building centers. Most are attached by driving finish nails through the top of the cap and into the post. Others (photo below) are screwed into the post top with preinstalled screws.

Building & Installing Porch Railings

A porch railing not only provides security from falls, but can also make an important contribution to the visual appeal of a porch. The basic components of a railing are the railing posts, the bottom and top rails, the balusters, and optional cap rails. It is usually easiest to assemble the railing before you install it. Check local codes before designing a railing. In most cases, railings should be at least 36" high, and the spaces between balusters and between the porch floor and bottom rail should be no more than 4".

Everything You Need:

Tools: basic hand tools, speed square, ratchet and socket set.

Materials: plywood, framing lumber, post caps, milled rail caps, lag screws, screws, siding rails.

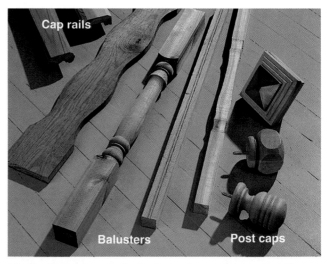

Railing components include decorative cap rails that are grooved to fit over a 2 × 2 railing; balusters, sometimes called spindles, that range from plain 2 × 2s to ornate millwork; and decorative post caps.

How to Build & Install a Porch Railing

1 Make a base plate for the railing post from two pieces of ¾"-thick plywood. Cut the plywood pieces to match the finished size of the post, including any post wrap boards (page 142). Stack the pieces together and fasten with 1¼" screws and construction adhesive. Do not put screws in corners.

2 Cut the post to finished height (usually 38" to 40"), allowing for the thickness of the base plate. If you plan to wrap the post, use scrap lumber the same thickness as the wrap boards to center the post on the base plate. Attach the base plate to the bottom of the post with three counterbored ⅜" × 4" lag screws.

3 Set the post in position on the porch floor. Drive ⅜" × 4" lag screws through pilot holes at each corner of the base plate, and into the porch floor.

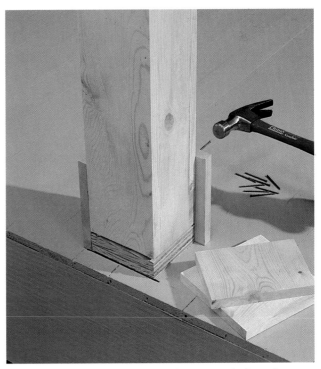

4 Cut and install wrap boards to match the other porch posts (page 142). Also cut and install collar boards to cover the edges of the base plate and the bottom of the post.

(continued next page)

5 Cut the top, bottom, and cap rails to length, using a circular saw. The bottom rail will have to be cut shorter than top and cap rails if you have installed collars at the post base. For our railings, we used 2 × 4s for the bottom rails, 2 × 2s for the top rails, and milled cap rails installed over the top rails.

6 Mark baluster layout onto the top and bottom rails—make sure balusters will be no more than 4" apart when installed. Drill ⅛"-diameter pilot holes all the way through the rails at the baluster layout mark, centering the holes from side to side.

7 If cutting your own balusters, clamp, measure, and cut the balusters to finished length. To save time and ensure uniform length, gang-cut the balusters, using a circular saw.

Spacing block

Shims

8 Lay the top and bottom rails on a worksurface. Use shims to support the balusters and top rail so they align with the center of the bottom rail. Attach the balusters to the rails by driving 2½" deck screws through the pilot holes in the rails, and into the baluster ends. TIP: Use a spacing block to keep balusters from shifting when they are attached.

9 Insert railing assembly between posts, using blocks to support it at the desired height. Drill ⅛" angled holes through the ends of the top and bottom rails and into the posts. Toenail railings to posts with 8d casing nails.

10 Slip the cap rail over the top rail. Attach the cap rail by driving 2½" deck screws at 18" intervals, up through the top rail and into the cap rail.

Variation: How to Install a Porch Rail Between Round Posts

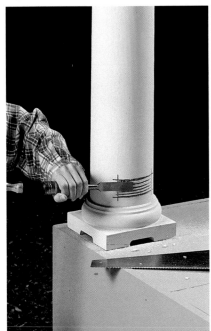

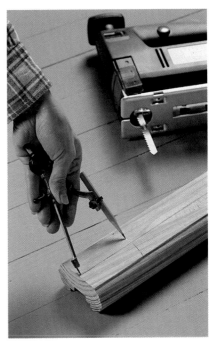

1 Assemble railings and balusters (page 150). Position the assembly between the posts. Mark the top and bottom of each rail onto the posts. Cut along the lines with a handsaw, then use a chisel to remove wood between the cuts.

2 Install the railing assembly by inserting the rail ends into the post notches. Toenail the rail ends to the posts.

3 Set a compass to the post radius, then scribe an arc on each end of a cap rail. The distance between the arcs should equal the distance between posts. Cut out the arcs with a jig saw. Attach cap rail to top rail (step 10, above).

Wooden porch steps are easier to build and less expensive than concrete steps. With paint and the appropriate railing design, they can also be made to match the look of the porch.

Building Wood Porch Steps

Wood porch steps consist of three basic parts. Stringers, usually made from 2 × 12s, provide the framework and support for the steps. Treads, made from 2 × 12s or pairs of 2 × 6s, are the stepping surfaces. Risers, usually made from 2 × 8s, are the vertical boards at the back of each step.

If you are replacing steps, use the dimensions of the old steps as a guide. If not, you will need to create a plan for the new steps, which takes a little bit of math and a little bit of trial and error (see next page).

In addition to the step dimensions, consider style issues when designing steps, and if there are two or more steps, include a step railing that matches your existing porch railing.

Porch steps should be at least 3 ft. wide—try to match the width of the sidewalk at the base of the step area. To inhibit warping and provide better support, use three stringers, not two. Wood steps do not require footings. Simply attach them to the rim joist or apron of the porch, and anchor them to the sidewalk at the base of the first step.

Tips for Building Steps

Remove the old steps, if any, from the project area. With the old steps removed, you can better evaluate the structure of the porch to make a plan for anchoring the new steps. If possible, attach the step stringers directly to the porch rim joist or to the apron. Also evaluate the condition of the sidewalk to make sure it is strong enough to support the steps.

Make a detailed plan for the steps, keeping in mind that each step should be 10" to 12" deep, with riser height between 6" and 8". Make sure the planned steps conform to the required overall rise and run (photos, below).

How to Measure Rise & Run

1 Attach a mason's string to the porch floor. Drive a stake where you want the base of the bottom step to fall. Attach the other end of the string to the stake, and use a line level to level it. Measure the length of the string—this distance is the overall run of the steps.

2 Measure down from the string to the bottom of the stake to determine the overall height, or *rise*, of the steps. Divide the overall rise by the estimated number of steps. The rise of each step should be between 6" and 8". For example, if the overall rise is 21" and you plan to build three steps, the rise of each step would be 7". It is very important that all risers and all treads are uniform in size.

How to Build Porch Steps

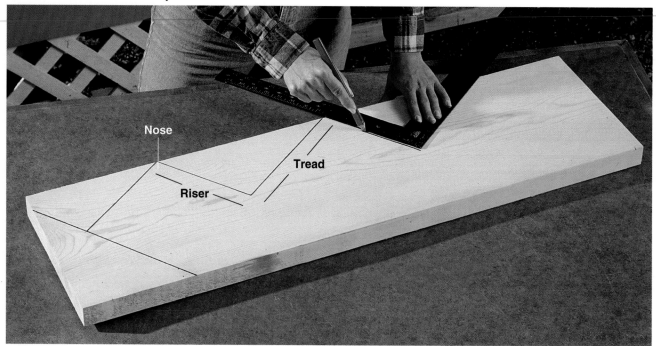

Nose

Tread

Riser

1 Mark the step layout (page 153) onto a board—usually a 2 × 12—to make the step stringers. Use a carpenter's square with the rise distance and run distance each noted on a leg of the square. Lay out the stringer so the "nose" areas where the tread and riser meet on each step fall at the same edge of the board. Check all angles with the square to make sure they are right angles.

2 Cut out the stringer, using a circular saw for straight cuts, and finishing the cuts with a handsaw where cuts meet at inside corners. Use this stringer as a template for laying out and cutting two more stringers. Mark and trim off the thickness of one tread (1½") at the bottom of each stringer so the rise of the bottom step will equal the rise of the other steps.

3 Attach evenly spaced metal angle brackets to the porch rim joist or apron, making sure the brackets are perpendicular to the ground. Position the stringers inside the angle brackets so the top of each stringer is 1½" below the top of the porch floor. Attach the stringers to the angle brackets with joist-hanger nails.

154

4 Measure the distances between the stringer tops, then use these measurements to cut two 2 × 4 cleats to length. Making sure stringers are square to the rim joist, attach the cleats to the concrete between the bases of the stringers, using self-tapping concrete screws driven into pilot holes.

5 Cut posts for the step railing to height, and attach them to the outside face of an outer stringer using 6" carriage bolts. Use a level to set the posts so they are plumb, then clamp them in position while you drill guide holes for the carriage bolts through the posts and the stringer. Drive the carriage bolts through the guide holes and secure each with a washer and nut on the inside face of the stringer.

6 Cut the risers and treads to length, using a circular saw. TIP: A 1" to 2" overhang at each outer stringer creates more attractive steps. Notch the treads for the top and bottom steps to fit around the posts, using a jig saw. Attach the risers to the vertical edges of the stringers, using 2½" deck screws, then attach the treads.

(continued next page)

7 Mark a 2 × 4 to use for the lower railing (see pages 148 to 151 for information on building railings). To mark the 2 × 4 for cutting, lay it on the steps flush against the posts, then mark a cutting line on the 2 × 4 at each inside post edge. Use the 2 × 4 as a template for marking cutting lines on the top and cap rails.

8 Set the blade of a circular saw to match the angle of the cutting marks on the 2 × 4, and gang the 2 × 4 with a 2 × 2 for the top railing and a piece of cap rail for cutting. Gang-cut the rails at the cutting lines.

9 Attach the bottom rail to the posts with deck screws driven toenail style through pilot holes and into the inside faces of the posts. The bottom of the rail should be level with the noses of the steps. Attach the 2 × 2 top rail so it is parallel to the bottom rail, 2" down from the finished post height.

10 Hold a 2 × 2 flush against a post so the ends extend past the top and bottom rails. Mark cutting lines on the 2 × 2 at the bottom edge of the top rail, and the top edge of the bottom rail. Use the 2 × 2 as a template for marking cutting lines on all railing balusters.

11 Mark layout lines for the balusters on the top and bottom rails, spacing the balusters no more than 4" apart. Drill ⅛" holes in the center of the top rail at baluster locations, then drive 2½" deck screws into the baluster ends. Toenail balusters to the bottom rail with 8d finish nails. Attach the cap rail (page 151).

12 Install a bottom rail, top rail, and cap rail in a horizontal position between the railing post at the top step and the end post for the porch railing. If the distance between posts is more than 4", install balusters between the rails.

13 Close off any open areas under the stringers with wood or lattice panels. First, attach nailing strips to the undersides of the outer stringer, set back far enough to create a recess for the wood or lattice. Cut a piece of wood or lattice to fit, and install. Attach decorative post caps (page 148) if desired, but first double-check the step post tops with a straightedge to make sure the tops follow the slope of the railing. Trim one or both post tops to height, if necessary.

Index

A

Aggregate,
  see: Exposed aggregate
Alignment and measuring tools, 11
Asphalt shingles, 140

B

Batter boards, 126
Beams,
  installing, 134-137
  materials, 123, 134-135
  minimum size requirement, 112
  wrapping, 123, 141-142
Bolts, types, 117
Bond-breaker for concrete, 49
Brick and block, 72-106
  basics, 78-81
  brick-paver step landing with
    planter option, 94-99
  brick veneer, 100-105
  building with, 92-93
  buttering, 84
  cleaning, 106-107
  cutting, 78-81
  decorative block screen, 73
  designing, 75
  double-wythe wall, 75, 77, 88-91
  end-support pillars, 75
  garden wall, 8, 73, 88-91
  header bricks, 88
  laying, 84-87
  marking, 79
  mortar throwing, 78, 82-83
  painting, 106-107
  patterns and strength, 76
  planning and design, 72, 74, 88
  reinforcing, 77, 87, 91, 93
  screen, decorative block, 73,
    92-93
  selecting, 8
  single-wythe wall, 75
  step landing, 73, 94-99
  tips for planning, 74-75
  types, 76
  water absorption rate, 78-79

Bricklayer's hammer, 11
Brick pavers,
  patterns, 76, 95
  variations for building with, 95
Brick paver step landing with
  planter, 94-97
Brickset, 10-11
Brick splitter, 81
Bricks tongs, 78
Brick veneer, 73, 100-105
Broomed finish on concrete, 30, 39
Building codes, 5-6, 8, 15, 34, 65-66,
  73, 75, 77, 93, 100, 112, 120
Building permits, 112, 118
Building site,
  grading, 110
  measuring slope, 110

C

Carpenter's square, 116
Carriage bolt, 117
Caulking backer rod, 57
Ceiling of porch, 145-146
Ceramic patio tile, 47
Circular saw to cut masonry, 10-11,
  79-80
Closure brick, 91
Cold chisel, 10-11
Combination corner block, 76, 85
Compactible gravel subbase, 75,
  84, 86
Concrete, 14-71
  basics, 16-19
  broomed finish, 30, 39
  building with, 32-33
  consistency, 16
  control joints, 48, 53
  curing, 29-31, 52
  demolishing old, 52
  edging, 29
  estimating coverage, 17
  evaluating concrete surfaces, 48
  exposed aggregate, 30-31,
    40-45
  finishing, 30-31
  forms, building, 24-25
  ingredients, 16
  maintaining, 70-71
  mixing, 16-18
  painting, 70
  placing, 26-29
  pouring concrete subbase for tile
    patio, 46, 49-52
  premixed products, 17
  ready-mix, 19
  reinforcement, 21, 25, 37, 43, 50,
    62, 64-65, 67, 69
  sealing, 30, 33, 39, 70-71
  site preparations, 14, 20-21
  steps, new, 60-65
  temperature range for pouring, 14
  walkways, 34-39
  wood forms, 24-25
Concrete mixer, power, 10, 16, 18
Connectors, types, 17
Control joints, 24, 28-29, 33, 37,
  39, 48, 53
  filling with caulk, 70
Cordless tools, 116
  and safety, 114
Corner blocks, 76, 87
Cornice, 143-144
Curing concrete, 29-31, 52

D

Deck joists,
  see: Joists
Demolishing old concrete, 62
Disposal, concrete products, 12-13
Double-wythe wall, 75, 77
  building, 88-91
Drainage, 21
Drill, types, 116
Drip-edge flashing, 125
Driveways, 9, 14

E

Efflorescence, 106
Egg splatter, removing, 106
Elevation drawing example, 113
Expose aggregate, 30
  creating exposed aggregate
    finish, 31, 44
  patio, 40-45
  sealing, 31, 45, 70-71

F

Fascia, 121, 138, 139, 144
  and soffits, 139, 145
  scalloped, 147
Fence columns, 6
Fiber additive for concrete
  reinforcement, 21
Fiber-reinforced concrete mix, 17
Fill, broken-up concrete for, 13, 62
Flashing,
  for porch deck, 125
  for roof, 123, 140
  types, 117
Floats,
  types, 11
  using, 29
Flooring nailer, 116, 131-132
Floor of porch,
  installing, 131-133
  materials, 122, 131
Floor plan example, 113

Footings, 66-69, 121
  and frost line, 112, 122, 127
  for brick and block, 67, 73-75
  for concrete steps, 62
  forms for, 25, 67-68
  frost line and footing depth, 23,
    66-67, 74, 112
  installing, 127-128
  pouring, 68
  reinforcement for, 67, 69
  tips for building, 67
  using earth as form for, 67
Forms,
  for footings, 25, 67-68
  for patio, 43
  for steps, 25, 63
  for walkway, 37
  how to build, 24-25
  permanent, material for, 43
  using earth as form for footing, 67
Framing square, 11
Frieze board, 121, 139, 144
Frost line and footing depth, 23,
  66-67, 74, 112

G
Gable, 143-144
  ornament, 147
Garage floor, 9
General-purpose concrete mix, 17
GFCI extension cord, 12
Grading a building site,
  measuring slope, 110
  when to regrade, 110
Groover, 11
Grouting tile, 58-59

H
Hammer drill, 116
Header bricks, 88
High-early premixed concrete, 17

I
Iron stains, removing, 106
Isolation boards,
  installing, 32, 37, 63, 68, 94
  purpose of, 21
Isolation joints, 21, 68
Ivy, removing, 106

J
Jack-on-jack, 76
J-bolt, 65, 117
  installing into concrete, 128
Jig saw, 116
Jointer, 11

Joist hanger, 117
  installing, 129
Joists,
  installing deck joists, 129-130
  materials, 122, 129
  recommended size, 112

L
Ladder safety tips, 13, 114-115
Lag screw, 117
Lap siding,
  cutaway diagram, 111
  removing, 111, 125, 140
Ledger boards,
  installing, 124-128, 134-135
  materials, 124
  minimum size requirement, 112
Lights in porch ceiling, 146
Line level, 11
Lumber types for outdoor projects,
  116-117

M
Magnesium float, 11, 31, 44
Masonry anchor bolt, 117
Masonry cutting blade, 11
Masonry grinding disk, 11
Masonry hoe, 10
Masonry tools, 10-11
Mason's chisel, 80
Mason's string, 11, 22
Maul, 11
Metal rebar, 21
  as tie rods in footings, 67, 69
  to reinforce masonry, 25, 37, 62,
    64, 69, 91
Metal reinforcing strips, 77, 87, 93
Mixer, power, 10, 16, 18
Molding, 146-147
Mortar,
  fortified, 82-83
  mixing and throwing, 82-83
Mortar bed, 83, 88
Mortar box, 10

N
Nailing cleat, 131
Nails, types, 117

P
Painting,
  brick and block, 106-107
  concrete, 70
Paint stains, removing, 106
Patio,
  concrete, 15

  design tips, 9, 41
  exposed aggregate, 40-45
  floating patio design, 41
  installing tiled patio, 46-59
  subbase for, 42
Patio enclosure,
  designs, 109
Permit, obtaining, 112
Plan drawings, 9
Plans, making, 110
Pod, 27
Porch,
  beams and trusses, installing,
    134-137
  building, 120-147
  ceiling, installing, 145-146
  cornice and gable, 143-144
  deck joists, installing, 129-130
  designs, 109, 118-119
  diagram of parts, 121
  finishing, 147
  floors, installing, 131-133
  ledger boards and posts,
    installing, 124-128
  railings, 148-151
  roof coverings, installing, 138-140
  slope for porch deck, 129
  soffits, installing, 145-146
  steps, 152-157
Post cap, 148
Post collar, 121, 142
Posts,
  anchor, 117
  installing, 124-128
  materials, 123-124
  minimum size requirement, 112
  round, and railings, 151
  wrapping, 123, 141-142
Post saddle, 136
Power drill for masonry, 10-11

Q
Quarry tile, 47

R
Rafters for porch roof, 123, 134
Rafter tie, 117
Railing,
  adding post cap, 148
  building for porches, 119, 148-151
  designs, 119
Ready-mix concrete, 19
Rebar,
  see: Metal bar
Reciprocating saw, 116
Reinforcement,
  fiber additive, 21
  for brick or block structures, 77,
    87, 91, 93

for concrete, 21, 25, 37, 43, 62, 64-65, 67, 69

for double-wythe brick wall, 77, 91

for exposed-aggregate patio, 43

for steps, 64-65

metal rebar, 21, 25, 37, 62, 64, 67, 69, 91

metal reinforcing strips, 77, 87, 93

wire mesh, 21, 25, 43

Rise and run of steps, measuring, 60, 153

Roof, 123

installing, 44-46

ledger, installing, 134-135

measuring slope, 110

shingles, 140

Round posts, installing railing, 151

Running bond, 76

S

Safety, 12-13, 20, 43, 62, 80, 114-115, 135

Sand mix concrete, 17

Saws, types, 47, 116

Scaffolding, 13

Screed board, 28, 43, 50-51, 62

curved, 35, 38

Screen, decorative block, 73, 92-93

Screws, types, 117

Sealer,

for concrete, 33, 65, 70-71

for exposed aggregate, 31, 45, 70-71

Seeding concrete,

see: Exposed aggregate

Shellstone tile, 47

Sidewalk, 9, 112

Siding,

see: Lap siding; Stucco

Single-wythe wall, 75

Site preparation,

concrete, 14, 20-21

laying out and excavating, 22-23

Slope,

adjusting for runoff, 23, 35, 40

for porch deck, 129

measuring, 21, 110

Smoke stains, removing, 106

Sod, removing, 13, 20, 23, 36, 42

Soffit,

and fascia boards, 139, 145

installing, 145-146

Spalling, 48

Speed square, 116

using, 116

Squares, 116

Stack bond, 76

Stains,

removing from brick and block, 106

removing from concrete, 70

Stair edger, 11

Steps,

brick-paver landing, 73, 94-99

building forms, 25, 63

building for porch, 119, 133, 152-157

concrete, 14-15, 60-65

designing, 60-61, 153

existing steps as porch foundation, 122, 124, 126, 130

fill for, 64

footings for, 62

gravel subbase for, 62

measuring rise and run, 60, 153

railings, 65

reinforcement for, 64-65

riser support, 63

types, 119, 152

Story pole, 11, 23, 32, 102

Stringer for steps, 154

Stucco,

cutaway diagram, 111

materials to complement, 109

removing stucco siding, 111

Subbase, compactible gravel, 21, 23, 32, 36, 40, 42, 62

T

Tamper, 10

Temperature range for pouring concrete, 14

3-4-5 triangle method, 22

Tie rods in footings, 67

Tile,

cutting, 46, 56

installing tile patio, 46-59

pouring concrete subbase for tile patio, 46, 49-52

sealing, 59

Tooling, 93

Tools needed,

finishing and curing concrete, 30

laying brick and block, 84-87

masonry, 10-11

mixing and throwing mortar, 82-83

placing concrete, 26, 28-29

sealing and maintaining concrete, 70-71

site preparation, 20

Trowel, pointing, 11

Trusses,

installing, 134-137

prebuilt, 121, 123

V

Variance, obtaining, 112

Veneer,

brick, 73, 100-105

W

Walkways, concrete, 7, 8-9, 15, 34-39

building, 36-39

crowning, 35

curved, spacers for, 9

designing, 34-35

reinforcement for, 37

slope line, 35

water runoff, 35

Waste material, disposal, 115

Wall cap, 76, 87

Walls, 9

double-wythe brick, 88-91

footings for, 15, 73

garden, brick and block, 73

poured concrete, 9

Wet saw, 47

Wire mesh for concrete reinforcement, 21, 25, 43

Wood float, 11, 29, 38, 44, 64

Wood forms,

see: Forms

Wrapping posts and beams, 123, 141-142

materials, 141

Wrenches, types, 116